D1832065

Science In Society
Some Perspectives

Science In Society
Some Perspectives

Edited by

Yash Pal
Ashok Jain
Subodh Mahanti

GYAN

GYAN PUBLISHING HOUSE
NEW DELHI-110 002
in collaboration with
NATIONAL INSTITUTE OF SCIENCE, TECHNOLOGY AND DEVELOPMENT STUDIES

Science in Society: Some Perspectives
(Science, Sociology, Development Issues)

ISBN 81-212-0458-5
Library of Congress Card No. 93-900317

©Authors

Published by
Gyan Publishing House
5, Ansari Road
New Delhi 110 002
in collaboration with
National Institute of Science, Technology
and Development Studies, New Delhi

All right reserved. No part of this book may be
reproduced in any manner without written permission

Printed at Goyal Offset Works, Delhi

Editors' Note

Scientists perceived accountability and their role vis-a-vis contemporary social realities have been undergoing continuous change. In recent years, within the scientific community , there is a growing realisation that scientists can no longer remain insulated to issues which are traditionally considered as outside the agenda of 'scientific investigation. Scientists domain of activities no longer have pre-defined exclusive contours. To play a meaningful role in social development, science has to be seen in the context of current social issues. This requires an ongoing dialogue between scientists and social scientists (including social activists). But mere realisation or acceptance of the need of such a dialogue is not enough. Scientists must actually engage in such dialogues. Scientists cannot often foresee that their professional knowledge can be useful in resolving issues like land reform, Babri Masjid--Ram Janam Bhoomi controversy, traditional skills versus modern technology, big dams versus small dams and so on, which apparently seem to be more in the domain of politics than of science. It is often believed that to engage in such issues means coming out of the professional interests and hence these issues are not worth their consideration. Consequently, scientists are often accused of being indifferent to social issues. This is not because the issues are 'un-scientific 'or the scientists are not willing to serve the social causes but it is because they grow up during their professional training under a particular science paradigm, where they are taught to break up phenomena into parts and sub-parts to make assumptions about everything else remaining constant, while a part is being studied. And only after analysing all the parts, they attempt to assemble them back into the whole. However, this very process of division and sub- division results into ever narrower specialisation. In recent years we see scientists failing to appreciate the inter-linkages among various scientific and social issues across time and space highlighted by environmental perspective and consequently instead of moving towards a holistic understanding, the environmental issues have been broken down into narrower and narrower specialisation.

The metaphor 'science in society' rather than 'science and society' enables us (as argued by Ashok Jain) to visualise a framework for understanding science, technology and society in a holistic manner. It marks a point of departure away from a reductionistic view of the world to rich diversity inherent in social processes. It is an attempt to unravel the complex interconnectivities between science, technology and society in their diverse manifestation. The 'science and society' agenda (which emerged outof concerns for peace, environmental degradation, impacts of industrialisation, the sustainable development vis-a-vis distributive aspects) had proposed interactions between scientists and social scientists but it presumed that both would remain confined to their respective domains in relation to their ultimate professional goals. This approach failed to initiate a serious and meaningful dialogue irrespective of its willingness to do so (within the framework of 'two cultures). This is because it did not challenge the exclusivity of science in true sense. Contrary to presumed character of science in the 'science and society' approach, science is no longer viewed as an independent entity or alienated from the society (as expounded by Yash Pal) in the 'science in society' approach. This approach attempts to: break down the barriers between specialisations; link modern science and technology with the traditional skills and knowledge; understand material reality in its totality; and locate society firmly within nature. It thus attempts to create a holistic knowledge.

The twenty-three articles that constitute the present volume, together do not present a cogent articulation of the various tenets of 'science in society 'approach, rather they form some kind of a collage. They present multiple perspectives. This volume should be seen as an extension of an attempt initiated by Yash Pal in the 77th Science Congress to highlight the need to move towards the realisation that we should come out of the old science and society approach. We are aware that to many, the selection of the articles will appear to be arbitrary. But then this volume presents the responses of some of the best minds of our society towards 'science in society' approach.

List of Contributors

1. **Rais Ahmed**, Former Vice-Chairman, University Grants Commission, C/o Indian Embassy in Mauritius
2. **Nitya Anand**, Ex-Director, Central Drug Research Institute, Lucknow 226 001.
3. **Amiya Kumar Bagchi**, Director, Centre for Studies in Social Sciences, 10, Lake Terrace, Calcutta 700 029.
4. **D. Balasubramanian**, Director, Centre for Cellular & Molecular Biology, Hyderabad 500 007.
5. **Pushpa M. Bhargava**, Ex-Director, Centre for Cellular & Molecular Biology, Hyderabad 500 007.
6. **V.G. Bhide**, Former Vice-Chancellor, University of Poona, Inter-University Consortium, Department of Physics, University of Poona, Pune 411 007.
7. **Chandana Chakrabarti**, Centre for Cellular & Molecular Biology, Hyderabad 500 007.
8. **D.P. Chattopadhyaya**, Project Director, Project of History of Indian Science, Philosophy and Culture (PHISPC), Camp Office, 25 Park Mansion, 57A Park Street, Calcutta 700 016.
9. **Harsh K. Gupta**, Director, National Geophysical Laboratory, Uppal Road, Hyderabad 500 007.
10. **Ashok Jain**, Director, National Institute of Science, Technology & Development Studies, New Delhi 110 012.

11. **K. Babu Joseph**, Professor and Head, Department of Physics, Cochin University of Science and Technology, Cochin 682 022.

12. **S.K. Khanna**, Vice-Chairman, University Grants Commission, Bahadurshah Zafar Marg, New Delhi 110 002

13. **Jayant V. Narlikar**, Director, Inter-University Centre for Astronomy and Astrophysics, Post Bag 4, Ganesh Khind, Pune 411 007.

14. **Yash Pal**, Former Chairman, University Grants Commission and President of the 77th Session of Indian Science Congress, Nuclear Science Centre JNU Campus, New Delhi 110 016.

15. **M.P. Parameswaran**, Kerala Sastra Sahitya Parishad, Parishad Bhavan, Trivandrum 695 029.

16. **Ashok Parthasarathi**, Additional Secretary Department of Scientific and Industrial Research Technology Bhavan, New Delhi 110 016.

17. **Vinod Raina**, Eklavya, E1/208, Arera Colony Bhopal 462 016.

18. **Y.S. Rajan**, Adviser, Department of Science & Technology, Government of India, New Delhi 110 016.

19. **T.B. Rajashekar**, Senior Scientific Officer National Centre for Science Information, Indian Institute of Science, Bangalore 560 012

20. **C.V. Seshadri**, Director, Shri A.M.M. Murugappa Chettair Research Centre (MCRC), Tharamani Madras 600 113

21. **Obaid Siddiqi**, Professor & Head, Molecular Biology Division, Tata Institute of Fundamental Research, Homi Bhabha Road, Bombay 400 005.

22. **B.V. Sreekantan**, Former Director, Tata Institute of Fundamental Research, Homi Bhabha Road, Bombay 400 005

23. **V. Sudarsen**, Professor, Department of Anthropology, University of Madras, Madras 600 005

24. **M.S. Swaminathan**, Honorary Director, Centre for Research on Sustainable Agriculture & Rural Development, 11, Rathna Nagar, Madras 600 005

21. Ghazi [illegible], Professor and Head, Molecular Biology Division, Tata Institute of Fundamental Research, Homi Bhabha Road, Bombay 400 005.

22. B.V. Sreekantan, Former Director, Tata Institute of Fundamental Research, Homi Bhabha Road, Bombay 400 005.

23. P. [illegible], Professor, Department of Anthropology, University of Madras, Madras 600 [illegible]

24. M.S. Swaminathan, Honorary Director, Centre for Research on Sustainable Agriculture and Rural Development, 3, Rathna Nagar, Madras 600 018.

Acknowledgements

'Science in Society' was the focal theme of Professor Yash Pal's Presidential Address at the 77th Annual Session of Indian Science Congress. In Cochin, the venue of the 77th Session of Indian Science Congress, Professor Yash Pal after seeing the overwhelming response from many scholars, thinkers and social activists to the theme decided to bring out a book capturing the various facets of 'science in society' approach. While the book is not a reproduction of actual presentations made at the Congress, the editors are deeply indebted to the Indian Science Congress — a unique and the largest gathering of Indian scientists for catalysing the book. Yash Pal's introduction to this volume is essentially based on his Presidential Address to the 77th Science Congress.

The editors are grateful to the contributors for readily agreeing to write articles based on their talks delivered at the Science Congress and on new themes for inclusion in this volume.

It is sad that the untimely death of Professor Sukhamoy Chakraborty has denied us of his article which he had so readily agreed to write. His article would have certainly enriched the contents of the book and added new dimension to the issues raised.

We thank the National Working Group on Patent Laws for giving permission to reproduce the article written by Dr. Nitya Anand. It was a generous gesture on the part of Late Shri Baldev Singh, then Secretary of Zaheer Science Foundation. to allow us to reproduce the Zaheer Memorial Lecture delivered by Professor Obaid

Siddiqi on the occasion of the 77th session of Indian Science Congress.

The editors are indebted to Shri Ajay Sheopuri and Shri A. Wahid of NISTADS for their help in preparing the camera-ready copy of the manuscript.

Finally, the editors would like to thank Shri B.P. Garg and Shri Satyendra Nath Sarkar of Gyan Publishing House for patiently tolerating our unsystematic working style.

CONTENTS

Science in Society : An Introduction

Yash Pal

The difference between "Science and Society" and "Science in Society" is very important for me. I have felt that much too often we consider science as an extraneous activity — a sub-culture which has been borrowed from outside, which has been useful in generating technology, mostly in other countries, and that technology then comes upon us and we are required to adjust our society, accordingly, to be in tune with its character. An impression is put across that science at best is a tool, an addendum, a source for providing means to a good life, a way of acquiring power and capabilities to influence others, sometimes to dominate others, and for solving some of the problems which arise, because we are growing in number, resources are small and we must improve our efficiency of controlling and exploiting nature. So, even when we recognise the importance of science, we only talk about "the impact of science and technology on society". However, I suggest that science should not be treated as an additional activity, not only as a tool, not as something which necessarily must arise in some better developed countries and come to us second hand; instead it should be treated as part of the culture of society, integral to our living, integral to our thinking, connected with our dreams, connected with concepts of ethics, beauty and spirituality and with the very deepest questions we ask in regard to "who are we, how are we related, where we came from, and where are we going." Certainly the tools which technology provides, technology which is becoming more and more soaked with science, are important. Certainly, means of production, goods and services, ways of looking after our land, increase of efficiency, modes of communication and transport to stay in touch with each other — all these are important. But, if we make science as a part of our culture they would all come, much more naturally, more creatively and more in tune with our living, not as an assault from outside.

Our origins, relatedness, joys and perspectives

Let me consider only one or two points in relation to this, what I would like to call cultural and spiritual aspects of science. This is something which gives me a great deal of personal sustenance. It is probably the basis of my personal ethics. It also gives me a great deal of joy. In addition it gives me a way of lending perspective to our present day problems, looking at them in a certain framework. I suggest that this could be a way of looking at ourselves which may free us from the tyranny of many of the prejudices which we have inherited from the past, in fact, even the tyranny of history — the so called "history".

We are all aware of the large number of problems which we have in the country itself, between countries and between people professing different faiths. We get agitated about things which may or may not have happened hundreds of years ago, questions like *Babri-Masjid, Ram-Janam Bhoomi* and other equally sensitive issues of this type. I think if we develop a truer, more reasonable, perhaps longer, perspective on our origins and development, many of these problems will appear to be either non-existent or will acquire a character which is not destructive.

How long have we been on this planet? Probably no more than 2 million years, as *homo-sapiens*. Two million years is a very small fraction of the total history of the planet, of 4.5 thousand million years, which is about a third of the life of the Universe. Out of those 2 million years, human beings have spent something like 94 percent of their life on this planet living in caves, wandering about the world, looking for food, not settled in villages, not even knowing agriculture. It is only during the last 6 percent of our life that we discovered agriculture and we started living in villages, towns and cities. If you talk of the period when great epics were written, a few thousand years ago, that is probably 0.001 percent of our presence on the planet.

And if you talk of modern scientific and technological civilisation, of industrialisation, theory of gravity or electromagnetism, the world of electronics, communication, new biology, aviation, atomic energy, space, computers etc., — all these are so recent, so very recent, probably 0.00001 of our time on this planet. So, in this great adventure of our survival and advancement, why do we pick on a few supposed events to make our lives hateful and troublesome? I am sure our troubles have other underlying, something proximate, causes which must be addressed, while simultaneously projecting a longer perspective. Our consciousness must be based on this perspective.

If we go further back to trace our connections with the whole of the cosmos, there is no question that our origins, at least origins of the stuff we are made of, date back to the very origin of the universe. It is quite clear, indeed indisputable, that the stuff we are made of, at one time or the other, was cooked in the middle of a star. We have stellar ancestry and we can trace our single lineage to the very beginning. This to me has a meaning of being connected to the whole, to what I see and what I do not see. The atoms within me are the same as the atoms within the sun and the stars, billions of stars. We do not know how many like us exist; certainly we have found that there is no form of life in our planetary system. Even if we are alone, it seems to be an amazing miracle that a, tiny part of the matter in the universe got involved in information-based self-replicating processes called life, finally evolving to a level of "conscious matter" which started asking questions about itself!

The fact that life is so rare in the solar system, and perhaps in the universe itself, and the fact that we humans arose only such a short time ago, is reason enough for me to develop a tremendous reverence for life and, in particular, for all we call human. We know now that like most of the planets in our solar system the atmosphere of this planet also started as being non-suitable for people like us. It was probably an atmosphere of carbon dioxide, and water vapour, may be with a bit of methane and ammonia. It is the miracle of life itself, early life which could live off those gases, which could flourish on those gases through development of photosynthesis, which could convert carbon dioxide into oxygen. And we owe it to growth of life on this planet itself that the planet became suitable for us human-beings who came later. That is enough of a miracle and that is enough of a reason for me to revere all life and to delight in the fact that we happened to come upon a planet where this was possible. And so it gives me not only joy and pleasure but also a sense of responsibility to look after this home and to care for it and consider everything on this planet as a relation. Whether it is a tree or a plant or a forest, the fish in the ocean or animals or people, we are all connected, we are part of the same web. We arose similarly, we are related, and understanding this is a joy, a bedrock on which to design our function. Our special privilege is that we can piece together the elements of this story, a privilege not given to all living things.

This story is not in the same category as mythology. Mythologies are beautiful. They are part of human living, part of human creative activity, search for values, search for connectivity, mixing of facts, imagination and poetry; indeed there cannot be a society without myths. We must distinguish, however, between mythology and a story put together with proven elements. The two are not incompatible. It is possible to enjoy why myths came to be. It is possible to enjoy the myths. But myths have their place and understanding has its place.

We should think of those beautiful pictures of this planet taken from space — the lonely blue globe against a vast darkness. Not that this was something which we didn't intellectually understand earlier. We knew that earth is a small planet around a big sun. But to be able to see ourselves that shimmering picture does give a feeling, firstly of our being so small and lonely in this wide universe and secondly of our being together on this planet for better or for worse. That ought to create for us a feeling of kinship, a feeling of belonging and understanding that in a sense we and the planet itself have together become one organism and I don't need any greater injunction for fellow-feeling, for being related, being dependent on each other, firstly having a common origin, and secondly, a common destiny, and that all our behaviour ought to be consistent with this realisation.

A certain widened kind of ecology should become the first religion of human-kind. A religion which does not have a brand name, and yet is consistent with the truth and sayings of all great religions which have been propounded, and whose wisdom is the same as the wisdom which has come through the holistic realisations of great seers of the past. If we begin to look at ourselves in this particular way, we will realise that many of the tensions which occur, many of the problems which we make for ourselves, many of the parochial fights which we have with each other have no deep meaning. We will see them in perspective. We will see them as little skirmishes, as fits of madness which, I hope, we can learn to avoid more easily than before. Perspective will bring us sanity and balance — also joy.

Non-alignment and innovation in science and technology

Of course, science and technology are connected with a whole lot of other things, connected with the whole way of living, producing goods and services, relations with other countries, other societies and so on. What is the connection between growing science in a society and this society's relationship to the outside world? Most people think that non-alignment and self-reliance are only political and sociological concepts. To me, basically, what non-alignment implies is that we shall stand on our own feet and take our own decisions. Non-alignment is a concept which was propounded in this country. But, I suggest that this concept has no meaning in the present-day world unless you also are innovative in science and technology. What does "being innovative in science and technology" mean? It means that you must create some new products, goods and services which are better than, or at least equal in quality to, those which anybody else can produce. It also means that your technology should specifically address your problems, irrespective of whether it is very simple or highly sophisticated. You cannot have a scientific and technological world or technological society which is dependent for everything it uses on innovations from abroad. If you have such a society it may appear to be a very glittering society for a while, but it cannot be non-aligned or self-reliant and the glitter won't last. It will be a dependent society. To the extent it is a dependent society, our cultural parameters will be largely defined by the science and technology coming from outside.

Technology or industry, for that matter, is not created in vacuum; it is created because of the urges of certain segments of the originating society. If it comes to you in the same form you will be forced to organise your own living in such a way that you conform to it and in addition, you may have to pay for it in other ways, either through your independence or through the riches created by your poor people, making them poorer still.

If we look at our country, we find two things. Yes, there is a great deal of science and technology in the country. I do believe that in many areas we have people who are equal to the best anywhere in the world. Yet, if we try to look for things which we have created, which are equal to or better than anybody else's and which bear our stamp, we will not find too many. Most of our industry is based on transfer of know-how. Most of the glittering goods we are beginning to find these days, whether they are television sets or VCRs or much of our telecommunication systems and computers, even toys, at least in metropolitan areas, are all copies of what is obtained abroad.

There are other things, of course, which are our own; there are many arts and crafts which are our own and better than anybody else's. There are the potters' things, and tie and dye techniques for fabrics and fashions. There is the famous *lota* which Charles Eames talked about. There are also in the technical areas many things which, I think, we have done rather well, but which we have not been able to produce in a way, in a quantity, so as to dominate any markets outside. And yet there is a lot of potential. One can see enormous number of bits and pieces all over the country where tremendous amount of innovativeness has been displayed, which, however, have been put aside on the excuse that we cannot meet time-targets, some trivial international standards are not met, or because of several other reasons, vested interests of various kinds, influence of multinationals, etc.

Any time, in any area of technology you find that you are coming to the threshold of excellent achievement, you find tremendous pressures coming in, which want you to put it aside, which accuse you of trying to re-invent the wheel. Nobody ever asks the question: why do the French make their telephone instruments themselves while the British had it already; and the Americans had it already and the Germans also make it themselves and Italians and the Japanese? Why don't they all buy the telephone instruments all from America or all from Japan?

I suggest that in science and technology you cannot imbibe the true spirit of the craft unless you do some things on your own, do your own inventing. It is necessary to make science and technology your own. You should ask the questions : Why are the Japanese, and the Germans and the French and the Americans all engaged in developing micro-processors and putting their own stamps? Why do they make the instruments somewhat different, in different countries, introducing minor changes, besides putting their brand names? Why must we compare ourselves with the technologically least advanced countries, and try to get everything from outside? In fact, you find that as soon as you are able to get to a situation where you are beginning to do something on your own there are various ways of killing it. Somebody will suddenly come with a loan offer coupled with a collaboration agreement, perhaps allowing you to use an Indian-foreign hypernated brand name. It is easy money for some industry, a few middlemen and everybody jumps at it; all the innovation which may have happened and could have fortified is killed.

There have been a few survivors, but these are exceptions to the rule. It would be a miracle if the recent bold experiment of mobilising indigenous science and engineering talent to design, build and market electronic switches of high performance is allowed to grow to its full potential. The blocking action is already on. It is understandable that foreign competitors would take all possible action to protect their own interest. But there are also vested interests within developing countries for whom a foreign collaboration is much more lucrative. We do not understand, deeply enough, that something which is beginning to grow here beyond a certain threshold is very precious; that if we kill it, we will kill all innovation in that sector for a long time to come. Innovation withers very fast if its products are not picked up, if it goes unrecognised. Countries which have grown a culture of innovation have done so by sustaining, by protecting it from outside; yes, critically examining it, not taking any nonsense and yet remaining conscious of its future potential, not only for meeting short-term requirements. Unless we have such a supportive attitude towards invention and innovations in the country it is unlikely that we will be able to stand in this tremendously competitive world where everybody is out to push his own things.

In a sense what I would like to emphasise is that the state of a society should not be measured only by taking snap-shots of its shop windows. The apparent prosperity reflected in them may be highly misleading. We should also examine the kind of processes which are operative in that society. Goods in the beautiful shop windows meant largely for the affluent members of society can be acquired by selling away the natural resources of your country, or bartering away the wealth created by the poor of your land. A country like India cannot be satisfied with this picture of development.

True development lies in the character of the processes which are operative in society and an important aspect of these processes is the existence of innovations. Are the people equipped and trained enough and do they have the right kind of attitude to stand on their feet? That is the only thing which will determine whether a society is self-reliant, non-aligned and truly free. That is the only way in which a society can choose its own directions for development, or emphasise, enhance and develop its own values. Being engaged in science and technology and creation of new science and technology is like a new "*Karma*". This is something we have to be doing in order to qualify for being intellectually whole, culturally self-dependent and really free. Without this, sustainable development, a self-reliant society or the true spirit of freedom are impossible goals.

Science education and science in education

We cannot do, of course, any science or technology nor for that matter anything else that is meaningful, unless we pay enough attention to education. We are supporting education in a fair way, but not adequately enough. We spend 2.5-3 percent of our total gross domestic product on education. About 15 percent of that is spent on higher education. Now, 15 percent of the total education resources on higher education is of the same order as most developing countries are spending. However, many of the developed countries, industrialised countries, are spending a larger fraction of their educational budget on higher education. This may be understandable because in many of the industrialised countries there are not so many young people; in fact they are having to close some of their schools. Our situation is very different. We have a whole lot of young people coming, and resources required for school education should be more in coming years. Nevertheless, we should not look at our expenditure in higher education only as a fraction of the resources which we spend on education, because expenditure on higher education has to be more related to creation of capabilities through which you will be competitive with the rest of the world. With expansion of transport and communication networks, hard sell and advertising of goods and images, the world has become a very competitive place in regard to goods and services, in regard to management, in regard to anything you do with the rest of the world. In your chosen areas, your capability must match theirs, not in terms of fraction of your per capita income but in absolute terms. Your people must be trained to the same levels, if not higher; otherwise there is no possibility of competing or standing on your own. So, the expenditure on higher education must be related to the expenditure on all economic activities in the country. It should be a certain fraction of what you spend on agriculture, industry, building of dams and roads, communication, defence, international relations etc.

I did an exercise recently, in connection with the work of the United Nations Advisory Committee for Science and Technology, to estimate the absolute amount which developing countries ought to be spending on science and technology education at higher education level. Being very conservative with respect to what is spent abroad, one came to a figure of something like 10 dollars per capita. (One has to talk of absolute amounts because equipment and facilities for science education cost the same amount per person). For India then we would need to allocate about 8,000 million dollars or, at the current exchange rate, Rs. 13,000 crores a year! We spend about 1,500 crores a year on all higher education. The difference is obvious. We may have to grow it slowly, but there is no question that the resources to be spent on science and technology education need to be increased by an order of magnitude.

We have been able to raise our expenditure in science and technology to a reasonable level, of the order of 1.2 percent of gross national product. But it is amazing that we spend less than half that amount on whole of higher education, including science and technology education. This is absolutely ridiculous and has no meaning if you want to sustain the very same science and technology you want to support. You cannot support science and technology research unless you are putting in money in the educational sector. Indeed the study I referred to earlier came with the conclusion that the amount of resources for S&T education and for doing S&T should be about the same. This would imply, even if we do not increase our S&T budget, that S&T education resources should be quadrupled!

Better equipped S&T teaching institutions would also increase the quality and quantity of their research. The advantage of intensifying research activity in the educational sector is that whenever you find anything new, a new process, a new way of looking at things, or fabricating equipment, there is an automatic transfer of this knowledge to the next generation. This won't happen if the investment is in laboratories where young people are not around. Much of it really goes with the people who make innovations. Therefore, it is important to realise firstly that we have to increase the resources in the education sector and secondly that a lot of research and development in the country has to be done in educational institutions if we want to make the process of growth sustainable.

In the present world the time between two innovations or changes of products and services is very short. Within 3-4 years a product changes its character, because some new innovations come in. The time which a student takes to go through a college or university is about 5 years. So, you cannot have the process of doing innovations and the process of learning being taken up serially. They must be done simultaneously. If you do it serially you will always be trying to catch up. That is another reason that the best of innovations have to go on and be done within educational institutions. This is a correction which we need to make very quickly and in fact some of the steps which we are trying to take in the University Grants Commission are in this direction.

Towards a planet assurance programme

There is a great deal of talk these days about the possibility of global warming. When we burn fossil fuels like coal, oil or wood and through other industrial activities, we generate gases like carbon dioxide and methane which, when they go into the atmosphere, act as thermal blankets. While Sun's energy can come in because it is at wavelengths which are not absorbed by these gases, some of the heat energy of the earth, being at wavelengths about twenty times longer, is absorbed by carbon dioxide and other gases in the air, causing a warming of the earth; radiation balance shifts to a slightly higher temperature. The level of carbon dioxide in the atmosphere has gone up considerably during last one hundred years because of industrial activity fuelled by fossil fuels. This is considered sudden because 100 years is too short a time compared to thousands of years, when it was fairly constant. It is not clear that this will definitely lead to a global warming, but some of the models indicate that in another 30-40 years the average temperature of this planet may rise by something like 2-3-4 degrees Celsius.

Now you may say, what is 2-3-4 degrees. After all we have such a large difference of temperature between summer and winter and we don't die, so what if the temperature is a little higher? It turns out that it makes a great deal of difference. With increase in temperature, the water column in the oceans will expand a bit, like mercury in a thermometer when you put it in your mouth. Just because of this, without the Antartica ice-shelf melting, a 5 degrees rise in year-round average temperature will lead to a sea level rise by a couple of metres. It may not be said that it is definitely proved that it will definitely happen, but it is likely. Then a large number of coastal towns will be under water, for example, a significant fraction of Bangladesh, Lakshadweep and our neighbouring country to the South, Maldives, will be almost completely under the sea. Millions of people may have to move. Bangladeshis would be crushed onto the borders of India. What will it do except leading to enormous disturbance, misery and conflict. Climate will change. Places which are supposed to be important granaries, such as North America, may become dust bowls. Some places, northern Siberia for example, may become marshes. Cyclones may become far more severe. Instead of only a few hundred people perishing in a cyclonic storm in India, many more may die. Floods may become much more severe. Earth has not known such a change for the last 5-10 thousand years. The average temperature of the planet has been constant within 2-3 degrees, one degree or so for the last 2000 years. We can certainly take and face such temperature changes quite easily if this happens over a 1000 years span. People slowly move to other places. They look for other ways of adjusting. But if it happens suddenly, then it will lead to a very difficult situation indeed.

So, what can we do about this? What about this carbon dioxide, for example? Can we stop putting carbon dioxide into the atmosphere? We cannot, because we do not know how to replace the vast amount of energy we now produce through burning of oil and coal and wood. Furthermore, we cannot even take this carbon dioxide and capture it in plants, because we are cutting trees. In fact, lot of carbon dioxide is being released in the atmosphere, because the forests which store a lot of carbon are also being destroyed.

It is instructive to look at the distribution of this "carbonisation" of the atmosphere. It seems that every US inhabitant puts in 5 tonnes of carbon into the atmosphere every year. This is what it takes to produce enough energy to maintain his style of living. In Europe it is about 3.5 tonnes, in China 0.2 tonnes and for an Indian it is about 0.15 tonnes per capita. The world average is about one tonne per capita per year. Therefore, it comes to a total of about 5 billion tonnes per year.

Let us now look at the future. China, India, Brazil and other countries are showing an industrial growth rate of the order of 8% per year. This would produce at least four doublings during the next forty years. If most of the energy for this were to come from fossil fuel burning, as is likely, the rate of carbonisation of the atmosphere will become three times, i.e., 15 billion tonnes per year, even if we assume that the first world will stay where it is! Whatever the uncertainties in the present theories, there is no question that this perspective will dangerously affect all human beings on the planet. The first world way of living when made applicable to a majority of the people in this world may be almost as devastating a nuclear winter. In this crisis also there will be no bystanders.

So, where do we go from here. It is true that most of the current problems, whether it is carbon dioxide or

CFCs, are due to those who use the most resources. India and other developing countries are yet far from being considered as global despoilers, no matter how inconsiderate we have been to our local environment.

We are way below the world average and there is no question that we have a lot of climbing to do. But our mode of development must be different and we should try to stay around the present world average even 20 years from now. We will need more energy efficiency, less waste and a lot of non-fossil sources. But it is clear that those who now use five times the world average and 30 times what we do just have to climb down and change their styles of living. But how do they come down? Have we ever been able to or is it ever conceivable to persuade an average American that he should give up his car, that he should not heat his house so much, should not throw away his junk, reuse his paper and plastic cups and in general that such a wasteful consumerist economy is not sustainable?

I have no doubt that if the world has to live in peace, we will have to converge towards a way of living which will not be so consumerist, which will not use the resources of the planet in such a way as it is doing now in the major industrialised countries. There would be pressures on us too; some brain-washing is already on. You would find that the western economists, western industrialists and western intellectuals are in great sympathy with anybody amongst us who shuns industrialisation and say that we should go back to villages and live simple lives. It is nice to live simple lives; I am not against that, but I resent people who themselves are putting 5 tonnes of carbon per capita in the atmosphere coming and lecturing me that I should not even use 0.5 tonnes because it is a common atmosphere; it is everybody's atmosphere, and it is ours as well as theirs.

We are coming to a stage where there cannot be islands of development. The world is tied together in this box we call the Earth. We can certainly, if we like, be consumerists, burn all the forests, so can Brazil and we can bring catastrophe to the whole world-if we also want to commit suicide ourselves. It is quite clear, if the arguments which I have presented earlier, are not enough, that living on this planet, all of us have a common destiny and a common responsibility for its hygiene. Even if we go away from the spectre of a nuclear war, which hopefully we might because of the increasing revulsion about where it might lead us to, we still have this major realisation in front of us that in the very process of what we have called growth and development, and industrialisation, we might be laying the ground work for a catastrophe that our children, certainly, and many of the young people will face, and that is not the heritage which we want to leave behind.

So, serious discussion and introspection have to take place in regard to styles of development, different roads to development, different forms of energy usage, questions of global security, questions, possibly, of a Global Public Sector which will ensure, for example, development, protection of biological diversity and care for the human commons. We probably have to reopen the debate in regard to fossil fuels of ordinary type versus nuclear fuels. The nuclear fuels don't produce any carbon dioxide, but they do produce radioactive waste.

Issues of security, war and peace, economic development, equity and disparity, styles of living and growth strategies, environment and ecology, even ethics and morality are all in an interacting cauldron. I believe we are beginning to understand some of these interactions, but a more general awareness is crucial to creating a *New Way of Thinking*. Everyone has some measure of responsibility towards the whole. While a grey uniformity is deadening for the human spirit and contrary to human aspirations, irresponsible consumerism and abject poverty cannot co-exist in a sustainable world. It is important that sociologists, economists, scientists, technologists, experts in international relations, and grass-root workers, join together to evolve the minimal parameters of what I would call a *Planet Assurance Programme*.

Ethics and law in an age of technological exuberance

As we increase in numbers, as we increase our capabilities, as we increase our powers and our exploitation of resources, we find that laws which we have created are not sufficient — national laws and international laws. They have not been designed to face this very fast explosive growth of science and technology and explosive growth of our powers and capabilities. We do not have laws on the books and we don't know how to treat the problems which would arise if, for example, we wanted to request Brazil not to destroy their tropical forests. We could argue that in wiping out their forests they would be destroying a common heritage of human-kind, because the biological diversity which exists in that tropical forest has come down to us over millions of years, and this contains enormous riches in the form of beautiful phrases of life which nature created over a long period of time and which all of us might need in our future to make sure that we can survive and prosper. How do you stop Brazil from destroying this human heritage if Brazilian farmers think, and Brazilian industry thinks, that they need to cut down the forests in order to provide energy, in order to provide farm land to their people? How will you compensate Brazilians in order to ensure that this wealth of the human-kind is preserved and how will you compensate them for not using that energy in building, even more, the concentration of carbon dioxide in the atmosphere because you are worried that this will change the climate? Will you force them? Will you occupy Brazil and say: you stop, we don't care, we don't want you to mess up with the future of the whole planet?

It is not only Brazil. It is also our tropical forests, in fact, tropical forests all over the world. Same way, how will you force an industrialised country where 20 percent of the people use 80 percent of the natural resources of the planet to start living more frugal lives! Can we say: no it is wrong, you shall not put 5 tonnes of carbon per person into the atmosphere; that there is a quota for messing up the atmosphere and the quota is only this much and if you go beyond that you have to pay a fee, that everybody is given a pollution licence, a pollution card and if you go beyond your allotted licence of pollution, then you have to pay a price, because using that money somebody else will reduce his pollution per capita. We don't have any laws, we don't have international laws, we have no way of negotiating except power, except using force and using force is not going to work. People don't even realise that it is illegal to do this, because it is not illegal at the moment.

So, we need to create new law and any new law which is created, of course, must have a relationship with new ethics. Ethics imply a code of behaviour. In its specificities, we may need to redefine the parameters of the code as we proceed. It was not unethical 500 or a thousand years ago to throw your muck into the Ganga. Not at all unethical, because Ganga had sufficient powers to purify it, because there was so little muck that Ganga would churn over and oxidise all the garbage you threw there. But certainly it is very unethical to throw all the garbage into the Ganga now, because Ganga *Maiya* at some places has become a sewer. It has lost the power to purify. And so ethics in many ways is a changing concept. If you want to define it specifically, there is absolute need to think of many issues connected with growth of science and technology, issues of scale, power and capability.

One specific area of concern, of course, is the question of personal privacy in the days of computers. Are there any laws? What kind of laws ought there be if your total information is in somebody's memory bank and he can use it in any way he likes? What kind of control should you have? Of course, there are some restrictions in this regard in some countries, but we still need to work at it without killing the possibility of further development. Similarly, a whole new field of ethics and law is needed in the area of new biology, particularly now that the whole human genome is going to be mapped. In another 20 years or so, the total code from A to Z of what is written in the human genome would have been read, all aspects which define a human being. The colour of your hair, the colour of your eyes, disposition of cells in our brain, and all the rest. This is very exciting, but also a little frightening. If we can add or take out little messages from here and there, what might we end up doing — and who decides, and how? There are many other dimensions to the issue of ethics in medicine and biology.

So, too much power through too much knowledge, necessarily means that we have to weigh, and we have to find ways in which it ought to be exercised and ways in which it ought not be exercised. I can quite understand if many people throw up their hands and say that knowledge itself is a sin, let us stop all activities in science and technology because it leads to these problems. This is impossible. You cannot bottle up minds. You cannot stop human thinking. That is the nature of human-kind. You will not stop it, but what you have to do is to block some of its applications. Block them by developing law and a code of ethics, because there is ample evidence that a whole lot of unethical and illegal things are not done very often. But we have not yet proceeded to develop appropriate ethics and laws in this direction, for what is coming in the future, even for what already exists.

Intellectual property and patents

Finally, there is an element which is connected with the whole future I believe, of the countries which have not yet made it to the starting line of science and technology, or which have just made it or are about to make it: this is related to the discussions on the intellectual property rights and patent laws. If somebody invents something, then, in order to encourage invention, we want to make sure that he gets some benefit out it. I suppose this was the origin of intellectual property rights. And so you take out a patent. You can sell your patent to a manufacturer. You tell him: this is my idea, I have invented it, you give me this much of compensation, you can manufacture and sell it. Or you can ask for a royalty on your patent or whatever you like. The patent laws have existed in most countries and they have been changing. Each country has had its own patent laws, at different points in time, depending on what it felt was beneficial to its interests. When Germans were far more advanced in technology, Britain resented their patent laws. It said, they were immoral, at least their type of patent laws. Now we have patent laws in our country too. But there are some aspects of our patent laws which allow us to take over those patents which we think are in public interest to operate. We can give compensation. If a certain drug has been patented by an outsider, it is possible for us to manufacture that or a similar drug through an entirely new process, our own process, and produce it in our country, because we do not have a product patent, we have a process patent. We do not recognise product patents.

If you find a new process, and make tetracyclines then it is new. Furthermore, we recognise patents for a period of 7-14 years. We think that is enough of a time to give advantage to the person who has patented it to get some returns for the effort he or she had put in. However, what is being now suggested, almost being forced on developing countries, some countries like India, is that we should accept the new patent laws which the Western countries have devised where the patents will be on products and not processes and, furthermore, the patents will have a lifetime which is of the order of 17-20 years. Now what are the implications of this ? If somebody produces one drug or a funny glass or a microphone, I can't produce that kind of drug, funny glass or microphone myself through any other technique or process. If I want to produce it in my country, I will have to buy his patent, but he can refuse. He himself is not obliged to produce it in my country. In that case the only way I can get that product is by importing from him at whatever price he charges, and this is valid for 20 years! Since most of the countries like India and developing countries have a very small fraction of the world patents now, it would essentially mean that from now on any new products which you want to produce in your country must necessarily be exactly those which have been invented outside and which belong to outsiders. Which means, in essence, that encouraging innovations in science and technology for producing goods and services in the country, is nonsense.

There is no use our having any science and technology for producing anything, because everything you get here will have to be essentially on the basis of knowhow transfer or working of the outsider's patents and his blue-prints. And this means the death of self-reliance, of non-alignment, of freedom, of trying to develop things our own way. I think we ought to resist it, no matter what the pressures. I do not know whether there is enough of a strength in our negotiating process or whether there will be too much of pressure put by the outsiders, threatening all manner of sanctions. But there is no question that this is a very critical item. People who are connected with the whole business of trying to create goods and services anew, and I am talking of scientists and technologists, understand it much more passionately than people who are in commerce, because the latter may look for a short-term advantage.

If you want to encourage a process of sustainable development in the country, there is no question that we must not accept these stipulations, because they would kill us at this stage. And we should not force others to accept this, those who have not yet arrived. But sure enough if you get to a certain stage where a reasonable fraction of the world patents are owned by you, then, along with all those countries which own a significant fraction of the patents, you can join a similar convention, because then there is fairness, you give as well as take. But in this situation it is completely unfair and it is planned essentially to ensure that those countries which had to develop, have already developed and those countries which are behind, shall never develop industrially and technologically. I don't think that is the kind of future which we would like to see or would like to compromise for.

I hope that I have been able to present, howsoever imperfectly, a few dimensions of some of the areas which I think lie in the realm of *science in society*. Science is important in its own right , as a part of human activity, but it is related to so may aspects of living and dealing with this world that unless we are aware of those and unless we work at them in a knowing way, we can't hope to build any kind of just and prosperous society. We can't talk of changing our economic policy and being decentralised, separate or different, unless we are self-reliant in science and technology. We can't have a self-determined, self-designed and self-propelled economy, centralised or decentralised, a socialist order or a beneficent market economy, or any other variety, unless we have the strength to stand free, to take our own decisions, and then to actually do what is best for us and what is best for our people and the world. In the process, we can take part in the marvellous, fantastic, unprecedented adventure of thinking, dreaming and creating.

The Physics-Astronomy Connection at the Most Fundamental Level

B.V. Sreekantan

Physics and Astronomy are two disciplines which for a long time in the history of science appeared to move in entirely opposite directions, physics in the direction of the ultra small — molecules, atoms, elementary particles and astronomy in the direction of the big and the distant — the planets, the stars, the galaxies, and the universe itself. However, in recent decades, the two fields have come together in an unexpected way as beautifully depicted in Fig. I, known as the Glashow's snake. Notice that the tail of the snake is in its mouth, the body is coiling round in a circle. This symbiosis and unification has tremendous implications for not only the two disciplines but also for science and philosophy in general.

Referring to Fig. I, we see 'Man' depicted in the scale range of 150-200 cm, somewhere in the middle of the bottom half of the circle. As we move to the left we meet

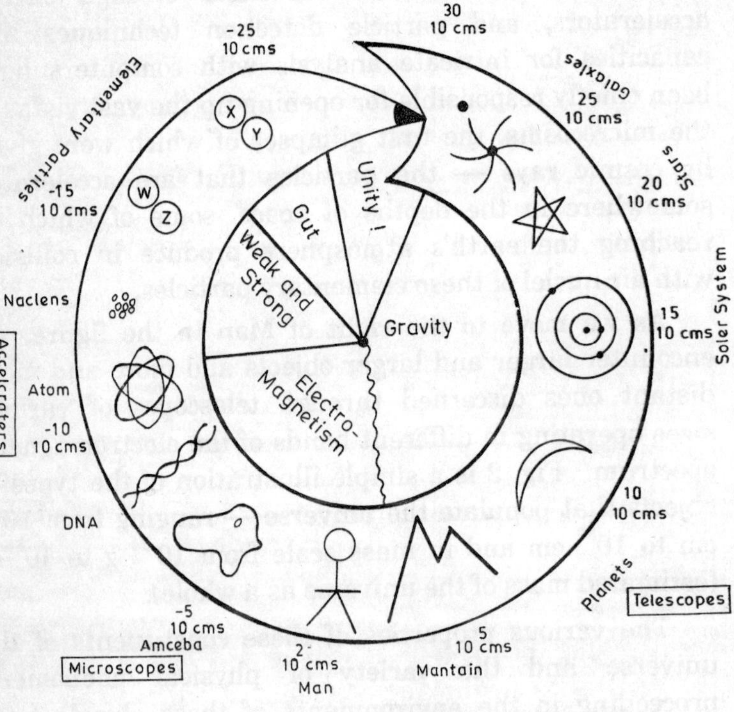

Fig. 1. GLASHOW'S SNAKE

objects of smaller and smaller dimensions — the *amoeba* — the single celled most elementary form of life, the DNA molecule, the molecules of various elements, the atoms, the nuclei comprising protons and neutrons and finally the species of elementary particles. The technological advances in the fields of high energy accelerators, and particle detection techniques and capacities for intricate analysis with computers have been chiefly responsible for opening up the vast vistas of the microcosms, the first glimpses of which were given by cosmic rays — the particles that are accelerated somewhere in the depths of space, some of which on reaching the earth's atmosphere produce in collision with air nuclei of these elementary particles.

As we move to the right of Man in the figure, we encounter larger and larger objects and more and more distant ones discerned through telescopes of various sizes operating in different bands of the electromagnetic spectrum. Fig. 2 is a simple illustration of the types of objects that populate the universe — ranging from 10^{-24} cm to 10^{30} cm and in mass scale from 10^{-24} g to 10^{56} g (estimated mass of the universe as a whole).

The various properties of these constituents of the universe and the variety of physical phenomena proceeding in the environments of these objects have been revealed to us not only through optical astronomy, but in recent decades also from radio, infrared, ultraviolet, X-ray, gamma and very high energy gamma ray astronomies. These new branches of astronomy became feasible with the technological developments in the fields of electronics and space technology.

One of the most significant findings of optical astronomical studies in the early decades of this century was the discovery that the universe is not static, but is

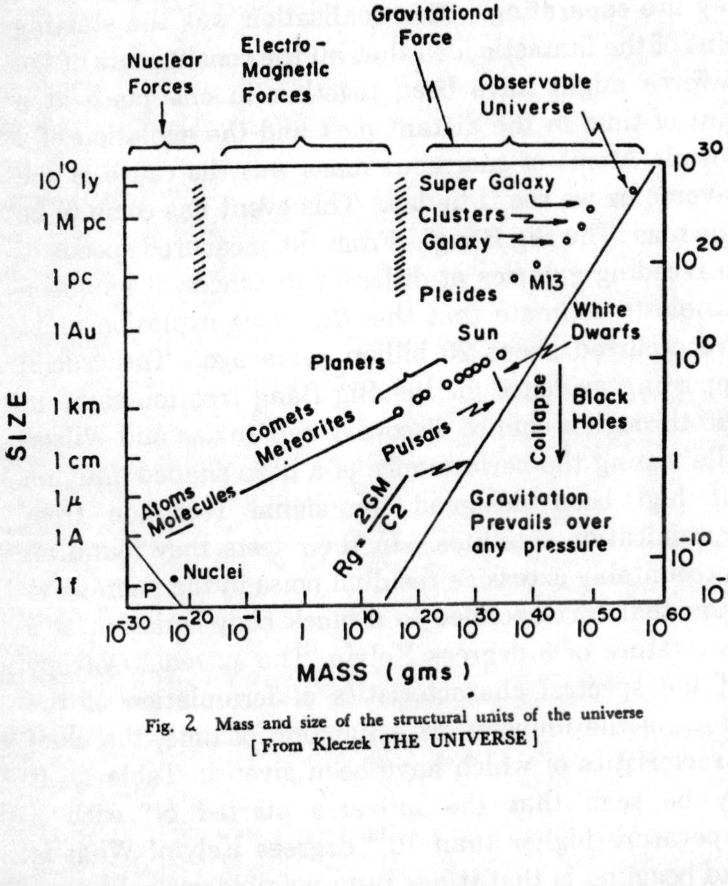

Fig. 2 Mass and size of the structural units of the universe
[From Kleczek THE UNIVERSE]

expanding — the galaxies are moving away from each other and surprisingly the greater the separation between the galaxies, the greater the velocity with which they are separating. This realisation was the starting point of the fantastic idea that all the constituents of the universe might have been together in one place at a point of time in the distant past and the explosion of a "Cosmic Atom" of enormous mass was the cause of the universe as we see it to-day. This event has come to be known as "The Big Bang". From the measured speeds of the receding galaxies at different distances, it has been possible to estimate that this Big Bang explosion must have occurred about 20 billion years ago. The crucial supporting evidence for the Big Bang creation came in 1965 through a chance discovery by Penzias and Wilson while testing the performance of a horn shaped antenna that had been designed for signal reception from communication satellites. In their tests they found an unaccountably excessive residual noise in the microwave region that corresponded to a black body emission at a temparature of 3 degrees Kelvin. The extreme isotropy and the spectral characteristics of formulation of the history of the universe as a function of time, the chief characteristics of which have been given in Table 1. It may be seen that the universe started off with a temperature higher than 10^{32} degrees Kelvin! What is mind boggling is that it has been possible not only to go back to the first few minutes of the Big Bang, but even to as small an interval of time as 10^{-43}seconds. The delineation of the happenings in the very early part of the universe is tied up with the developments that have taken place in the field of high energy physics.

Referring back to Glashow's Snake (Fig. 1), we see that basically four forces operate in nature — the

Table 1

Important Milestones in the History of the Universe

Cosmic Time	Epoch notation	Red Shift	Nature of the Phenomenan
0	Singularity	Infinity	Infinite density; Zero of time
10^{-43} Sc	Planck Time	10^{32}	Particle creation begins
10^{-6} Sc	Hadron Era	10^{13}	$P^{+}\bar{P}$ annihilation
10 Sc	Lepton Era	10^{10}	$e^{+}\bar{e}$ annihilation
2 mins.	Radiation Era	10^{9}	Nucleosynthesis of Helium, Deuterium
70,000 yrs	Matter Era	10^{4}	Matter dominates the universe
300,000 yrs	Decoupling	10^{3}	Universe becomes transparent to radiation
1.2×10^{9} yrs		10 - 30	Galaxies form
3×10^{9} yrs		5	Clustering of galaxies begins
4×10^{9} yrs			Formation of Milkyway Galaxy
4.1×10^{9} yrs			First Stars form
5×10^{9} yrs		3	Quasars born - Pop II Stars
10×10^{9} yrs		1	Pop I Stars form
15.2×10^{9} yrs			Parent Interstellar Clouds that gave rise to Solar System form
15.3×10^{9}			Collapsed Proto-Solar Nebula
15.4×10^{9}			Planets form
20×10^{9}			Homo Sapiens appear (Just 10^{5} years ago

gravitiational force which is the most important force in the context of large bodies and large distances; the electromagnetic force reigns supreme as far as phenomena associated with molecules and atoms are concerned. As we go to the nuclear and elementary particle domain, two more forces responsible for the spontaneous disintegration of particles (radioactive decay) the weak force and the strong force that holds the particles together inside the nucleus and also operates in the production of new particles in high energy collisions become the dominant ones.

Theoretical insights and experimental results with accelerators of higher and higher energy in the last three decades have led to some profound conclusions on these forces. In the ultimate analysis it turns out that the different forces are perhaps manifestations of one and the same force, the widely different features being connected with distance scales and energy transfers in the different phenomena observed. The scales of energy and distance of approach required for a full understanding of the behaviour of these forces are available only when we deal with high energy elementary particles and their interactions.

The electron and the proton were discovered as fundamental particles of nature in the very early part of this century through the study of discharge phenomena in low pressure gas tubes. The neutron, another fundamental constituent of all matter, was discovered in 1932 through the bombardment of α particles from a radioactive source on beryllium. Between 1930 and 1953, a series of new particles that lived only for a very short time were discovered in the analysis of cosmic radiation passing through the atmosphere. Among the particles discovered in cosmic radiation are positron (the anti-particles of electron), the mu-meson, the pi-meson, the k-meson, hyperons, and the $\Sigma^{\pm}$ and the cascade minus. These discoveries and the developments in the field of nuclear physics motivated the construction of higher and higher energy accelerators that became feasible with the advances in vacuum and electronic technologies. With the accelerator produced beams of protons and other particles it not only became possible to produce and study in detail the various elementary particles discovered in cosmic radiation but also to recognize the existence of extremely short lived particles, ($\sim 10^{-23}$Sc) known as 'resonances'. Very elegant phenomenological theories followed which resulted in the prediction of some more new particles with special properties, which were also discovered at the accelerators. However, a disturbing feature was that the number of elementary particles swelled into hundreds. It looked as if all the simplicity and elegance of nature, at the fundamental level was lost. To restore this simplicity and elegance of nature "the quark model" of hadrons was proposed which proved eminently successful. In this picture all the baryons and mesons are composites of more elementary entities called 'quarks' and their anti-particles, the 'antiqaurks'. There are six flavours of quarks (see Table 2) and in each flavour there are three varieties distinguished by their 'colour'. The quark-quark forces are mediated by a set of new particles called 'gluons' which are massless bosons. Each gluon also carries the colour charge. In parallel to Quantum Electrodynamics, the theory of quark-quark forces called Quantum Chromo Dynamics (QCD) has been developed.

Though a free quark has not been seen either in cosmic rays or at accelerator, experiments on the

Table 2 Quarks-Properties

Generation	Nomenclature (Flavour)	Electric Charge	Baryon Number	Spin	Strangeness Number	Charm Number	Mass (Gev)
I	up (u)	$-1/3^{e}$	1/3	1/2	0	0	0.336
	down (d)	$2/3^{e}$	1/3	1/2	0	0	0.338
II	Strange (s)	$-1/3^{e}$	1/3	1/2	1	0	0.54
	Charm (c)	$2/3^{e}$	1/3	1/2	0	1	1.5
III	Top (t)	$2/3^{e}$	1/3	1/2	0	0	Not seen yet
	Bottom (b)	$-1/3^{e}$	1/3	1/2	0	0	0.5

- All the quarks have corresponding anti-quarks with opposite Electric Charge, Baryon Number, Strangeness and Charm Numbers.
- Each Flavour of quark has 3 coloured charge states (red, green and blue)

  For example $\Delta^{-} = d_r\, d_g\, d_b$
- Baryons are combinations of three quarks of different colour; meons are combinations of quark and anti-quark.

Quark Structure of Baryons (Examples)

p	n	Λ^0	Σ^+	Σ^-	Ξ^0	Ξ^-	Ω^-	Δ^-
uud	*udd*	*uds*	*uus*	*dds*	*uss*	*dss*	*sss*	*ddd*

Quark Structure of Mesons (Examples)

π^-	K^+	Υ	D	F
$d\bar{u}$	$u\bar{s}$	$b\bar{b}$	$c\bar{u}$	$c\bar{s}$

scattering of electrons and neutrinos on protons have given full support to the quark structure of nucleons. An important property of the QCD force is that it depends on the momentum carried by the mediating particle, the gluon -- surprisingly the higher the momentum, weaker the force. As a result the quark behaves as a free particle at extremely short distances and is permanently confined within the particle since the force increases at larger distances.

While hadrons could be treated as composites of quarks, the leptons did not lend themselves for further division and had to be given the same status as quarks. Fortunately, the number of leptons, the light mass weakly interacting particles, has not increased. In addition to the electron, the muon and the neutrino and their anti-particles, only the τ-meson was discovered as a member of this family and for theoretical reasons a τ-neutrino had to be postulated, though not established experimentally yet.

So, at the fundamental level, we have moved from protons, neutrons, and electrons as constituents of all matters to quarks and leptons.

Among the four forces, the first unification, clearly perceived theoretically and established experimentally is the electro-weak unification according to which the difference between the electromagnetic force and the weak force disappears at an energy scale of 100 Gev corresponding to a distance scale of 10^{-13}cm. The behaviour of the strength of the strong force with energy transfer and that of the electro-weak force is shown in Figure 3. It is seen that there is a clear trend for the merger of the three forces at energy scale beyond 10^{15} Gev. This is too high an energy to be realised at accelerators even in the next century or in the subsequent ones. One of the crucial predictions of this Grand Unification Theory (GUT) is the decay of the proton, with a life time of the order of 10^{30} years several experiments including the one at KGF have been carried out on proton decay. While decay with a life time of 10^{30}years and into the most dominant modes predicted by the so-called standard theory have been ruled out, the possibility that the life time is longer than 10^{32} years and the decay modes distinctly different from the standard theory prediction are not so far ruled out. Yet another prediction of the GUT is the existence of massive magnetic monopoles. It is in this context that the particle physicists have become very much interested in the very early universe which turns out to be the highest energy accelerator that could ever be produced; and that was in existence 20 billion years ago.

The scenario of the early universe that has been formulated on the basis of many inputs from astronomy and elementary particle physics is as follows (Table 3): As we go back in time, the universe was smaller and smaller and hotter and hotter. When the universe was 3 minutes old, temperature, (which is to-day 3 degree Kelvin) was close to a billion degrees (10^9 degrees Kelvin) and as we go to still earlier times, for example around micro-second, the temperature should have been (10^{13} degrees Kelvin). This would correspond to individual particles having an energy of a billion electron volts. At still smaller time, say 10^{-12}second of Big Bang, the estimated temperature is 10^{16} degrees Kelvin and the corresponding energy of particles 10^{12}ev (Tev). Around 10^{-36} seconds the energy would be 10^{24} ev. The present extrapolation goes upto 10^{-43}seconds, when the temperature is 10^{32} degrees Kelvin and the energy of the particles 10^{28}ev. It is seen therefore that in the very

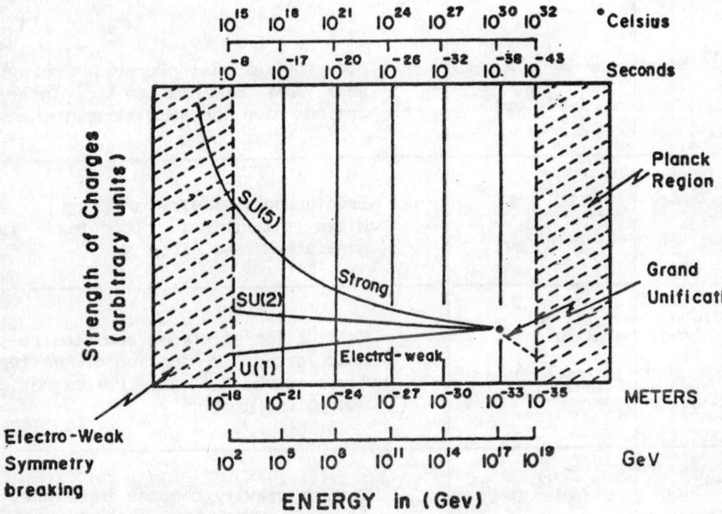

Fig. 3. The very early Universe and Grand Unification
[From M. Green SUPERSTRINGS]

Time From Big Bang	Temperature	Features of the Universe
10^{-6} Sc.	$> 10^{13}$ °K ($>10^{9}$ ev)	Non-gravitational properties change. Weak and Electromagnetic Interactions have the same strength.
10^{-13} Sc.	$> 10^{16}$ °K ($>10^{12}$ ev)	Spontaneous Symmetry breaking Higgs Mechanism operates to generate masses of $W^{\pm}$, Z^{0}
10^{-36} Sc.	$> 10^{28}$ °K ($>10^{24}$ ev)	Unification of strong and Electro-Weak forces. Production of massive lepto-quarks $X, \bar{X}$, massive magnetic monopoles, exotics.
10^{-43} Sc.	$>10^{32}$ °K ($>10^{28}$ ev)	Quantum gravity becomes important. No good theories yet to make predictions.

Table 3
Happenings in the Universe in the First Microsecond

early time of the universe, 10^{-43} seconds to 10^{-36}seconds, the energy of the particles (essentially quarks) could have been sufficiently high to bring about the unification of the three forces — strong, weak and electromagnetic. The feasibility of such an extrapolation is connected with the fact that the quark is a point particle (10^{-18} cm). The limit of 10^{-43}seconds is because of the lack of a clear quantum gravity theory. The evidence for this hot super-dense state would be the production of exotic particles including the "magnetic monopoles" of very heavy mass (10^{15}Gev/c 2). Some of these could have survived upto the present time. Search for exotic particles and magnetic monopoles is on in many laboratories in the world.

This high temperature high density scenario of the early universe has proved to be of great value in explaining some intriguing aspects of the universe like the dominance of radiation over matter and matter over anti-matter which have defied an explanation so far. Thus, physics and astronomy have become disciplines which are inextricably intertwined at the most fundamental levels.

For those interested in knowing more about this exciting development, I would like to refer to an article of mine "*The First Moments of the Universe*" in Transactions of the Bose Research Institute, Vol. 50, No. 1, 1987, and the references therein.

Symmetry, Asymmetry, Beauty and Science

Pushpa M. Bhargava
Chandana Chakrabarti

There are many ways of looking at symmetry and asymmetry from the merely descriptive to the mathematical, the philosophical and the abstract. One could look at symmetry in biological systems, in chemical systems (such as molecules), and in physical systems. One could thus analyse the nature, origin, structure and purpose of symmetry in a system and the consequences of asymmetry in the system. For example, externally, our body appears quite symmetrical; inside it is symmetrical in many ways but with notable exceptions, say in regard to the heart. In biology, therefore, one could look at the advantages and disadvantages of symmetry (or of the lack of it) from the Darwinian evolutionary point of view. And, one could study the relationship between art and symmetry-asymmetry. In all such studies, one assumes symmetry to mean "exact likeness in size, shape, form, etc., between the opposite sides of something", and asymmetry to mean the lack of such exact likeness; symmetry and asymmetry are thus juxtaposed as two non-overlapping sub-sets.

The three purposes or objectives of this article are to resurrect, what we believe is the original meaning of symmetry — the way Greeks understood it. (Symmetry is a Greek term and a Greek idea.) This "other" meaning of symmetry implies, according to various dictionaries and encyclopedias; correspondence of parts, regularity, conformity, balance, proportionality, equilibrium, order, harmony, beauty of form, shapeliness, beauty resulting from right proportion between parts of a body and the whole, congruity, rhythm, equipoise, stability, and evenness of structure. It is this definition of symmetry that immediately relates it to beauty. To establish such a relationship is the second objective of this article. The third objective is to show that beauty and science are closely related: in fact, so closely related that they could be thought of as two sides of the same coin.

Asymmetry is the antonym of symmetry only in one respect, that is, in respect of one meaning of symmetry: the exact likeness in size, shape or form, etc., between the opposite sides of something. Therefore, an object which is asymmetrical in the sense that its sides are not mirror images, could still perfectly well satisfy the second definition of symmetry given above, it could be harmonious and beautiful.

Symmetry and beauty

We have used many terms above to try to describe what we believe the Greeks had in mind when they invented the term "symmetry". All these terms involve a degree of value judgement. How would you, for example, define beauty or shapeliness or harmony? Yet it would be obvious that the elements of symmetry, as originally conceived by the Greeks, are closely related to the elements that we intuitively recognize as being descriptive of beauty. The distinguished physicist, W. Heisenberg, defined beauty as " the proper conformity of the parts to one another and to the whole" — a definition virtually identical with one of the attributes mentioned above of symmetry. An object which is asymmetrical in the sense that it is not comprised of mirror images can, therefore, be just as beautiful as an object which is symmetrical in terms of the generally accepted and commonly conceived notion of symmetry today. From the point of view of beauty, therefore, it should not be a question of symmetry versus asymmetry. It would be a question of the presence or absence of a logic in form, structure, composition, colour, texture, or other relationships: a logic that generates harmony and is compatible with the intuitive sense of beauty. Symmetry, by its original definition, *is* an expression of such logic; asymmetry *can be* so. To understand this logic, we need to look at the anatomy of beauty in a way that parts are still able to describe the whole. An extremely difficult task but we will attempt it even at the risk of being trite.

Beauty is related to science

The renowned 12th century Sanskrit poet, Jayadeva, in his immortal classic, "*Geeta Govinda*" extols the beauty of Radha in these words:

Thy *brow like smooth Bandhuka-leaves*; thy cheek
which the dark-tinted Madhuk's velvet shows;
Thy *long-lashed Lotus eyes*, lustrous and meek;
Thy *nose a Tila-bud; thy teeth like rows*
Of *Kunda-petals....*
....thy breasts like jasper-cups,

....thine eyes like planets;
Thy fragrant hair, thy stately neck,
Thy queenly sumptuous head;
Thy soft small feet, thy perfect lips,
Thy *teeth like jasmine petals*,....

The underlined similes used by Jayadeva to describe Radha are exquisite. And they are exquisite because they all refer to objects created by Nature.

Indeed all artists, be they poets, painters or sculptors, have been inspired through history by creations of Nature, be it the beauty of a woman or of a flower, or of the myriad other things that Nature has to offer with its infinite variety of form and function. The horses of Hussain seem both of the Earth and of Heaven. Renoir chose Gabriel as his favourite model, for her skin "took the light so beautifully". And all objects made by man that we intuitively consider beautiful are those which blend with Nature and are in harmony with it.

What, then, is it about Nature that generates so much beauty in the minds of those who behold it? What is it that is common between beauty painted on the canvas of time by nature and that created by man? The common element is *order*.

Nature generates order following what we call 'natural laws', such as the laws of science — of physics, chemistry and mathematics. Beauty arises in Nature out of the simple, universal laws of science, just as the laws of science arise out of beauty in Nature. The perception and recognition of the laws of science by man is nothing more than an expression of his ability to perceive beauty. This is then the seminal relationship between beauty and science: they are two sides of the same coin — two inter-related, interdependent manifestations of Nature.

There are abundant examples of Nature being orderly, and of beauty — both natural and man-made — following mathematical rules, mathematics being the ultimate abstraction of all sciences. Take, for example, the plant world, which has been such an important part of the folklore of beauty. If you count the number of leaves on successive branches of a plant or a tree, starting from the bottom, the numbers you are likely to obtain would not be random. They would fall in a series — a Fibonacci series. In such a series, the next number is always the sum of the preceding two numbers: such as 1,1,2,3,5,8,13,21,34,55, and so on. The arrangement of the whorls on a pinecone or a pineapple, or of petals on a sunflower, also follows a sequence of Fibonacci series. Similarly, the patterns we witness on the wings of a butterfly, on the cuticle of an insect, or in the web of a spider, follow distinct mathematical logic. Isn't it incredible that so much order exists amidst apparent chaos.

The aesthetic value of the golden ratio, 1:1.618, has been recognised for long by painters and architects. This ratio has many interesting properties. For example,

$$\frac{1}{1.618} = 0.618.....$$

$$\textit{And}\ \frac{(\sqrt{5}+1)}{2} = 1.618,\ \textit{while}\ \frac{(\sqrt{5}-1)}{2} = 0.618$$

In a Pentagon (Fig.1),

$$\frac{b}{a} = 1.618$$

$$\frac{a+b}{b} = 1.618$$

The ratio occurs repeatedly in the other regular solids as well. And it abounds in the molecular

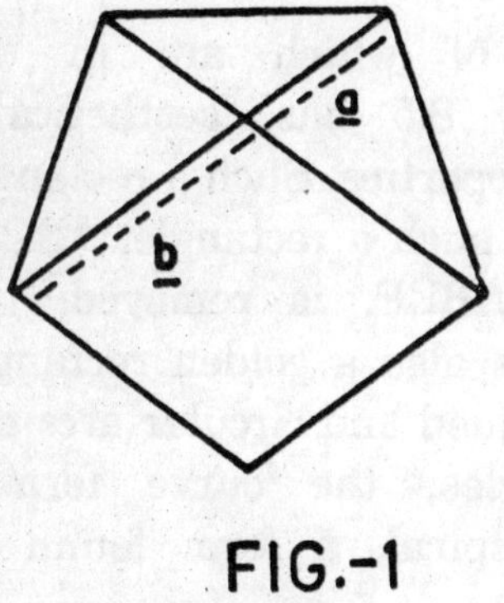

FIG.-1

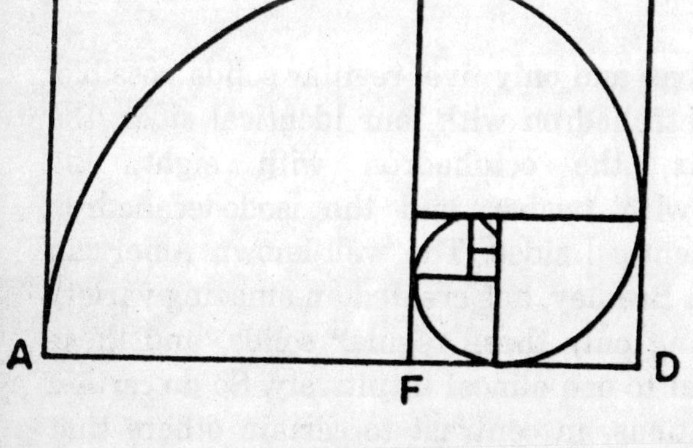

FIG.-2

architecture of DNA, the genetic material, too! Sections **a** and **b** of line, xy, so that b/a = 1.618 and a+b/b =1.618, are called the golden sections, and have played an important role in both ancient and modern artistic and architectural design.

1.618 = approx. 8/5, and 0.618 = approx 3/5

A rectangle, the sides of which are in the approximate ratio of 3:5 or 8:5 has, aesthetically speaking, the most pleasing properties. Such a rectangle is called a golden rectangle. If such a rectangle, ABCD, is drawn and the square, ABEF, is removed, the remaining rectangle, ECDF, is also a golden rectangle (Fig.2). If this process is continued and circular arcs are drawn through the rectangles, the curve formed approximates a logarithmic spiral, a form found in Nature.

Indeed, mathematical relationships such as the above are the basis of the theory of dynamic symmetry which has been central to the living forms in nature on the one hand and to the arts — the handiwork of man — on the other.

There are five, and only five, regular solids possible: they are, the tetrahedron with four identical sides, the cube with six, the octahedron with eight, the dodecahedron with twelve, and the isododecahedron with twenty identical sides. The well-known American painter, Morton Bradley, has created an amazing variety in painting using only these regular solids, and these paintings appeal to one almost intuitively. So do certain colour combinations, in contrast to certain others that just do not seem to click.

The visible spectrum, as is obtained, for example, in rainbow or through the use of a prism, consists of light of different wave-lengths or frequencies. On the increasing wave-length scale, one goes from red through orange, yellow, green blue and indigo, to violet. This is also the order in which we perceive colours as warm and /or cool. Red and yellow are universally regarded as warm colours, whereas blue and violet are regarded as cool colours. The concept of harmony and disharmony in colour composition is at least partly based on the relative position of colour in the visible spectrum.

Human eyes cannot easily distinguish between objects which are less than, say, a tenth of a millimetre apart. Nature's relentless quest for order and beauty leads it to generate as much beauty in tiny spaces that eyes cannot perceive, as over larger spaces. To be able to see what exists in nature in space as small as a thousand of a millimetre, or even smaller, one has to use a microscope. The most powerful of the known microscopes is the electron microscope which can distinguish between two points as close as two millionths of a millimetre apart. If one looks through the eyes of an electron microscope at some of the art-works of Nature, it becomes obvious that Nature is orderly and beautiful all the way down to the minutest detail that is totally invisible to the human eye. All forms and structures we see in Nature — animate or inanimate, from rocks to man — are the result of the use of scientific laws by Nature: laws, the ultimate language of which is mathematics. Beauty and science again come out to be inseparable — even where our five senses fail to perceive it unaided.

The perception of beauty of man must have given him an advantage during evolution and competition with other species for existence. In other words, man may not have developed to the extent he has, and come to be at the top of the evolutionary hierarchy as he is today, if he had not learnt to appreciate the beauty of natural objects and phenomena. Learning science — that is, understanding natural phenomena — has also been of advantage to man. He must have found that out early during evolution. And it is only through intuitive recognition and appreciation of beauty that he would have come to recognize and be motivated to understand the order which underlies all sciences.

As Poincare has said:

> "The Scientist does not study Nature because it is useful to do so. He studies it because he takes pleasure in it; and he takes pleasure in it because it is beautiful. If Nature were not beautiful, it would not be worth knowing and life would not be worth living.... I mean the intimate beauty which comes from the harmonious order of its parts and which a pure intelligence can grasp."

As we would, therefore, expect, man's perception of beauty has evolved over the ages, as has his understanding of science, that is, of Nature. An evidence of this is seen in the shift in the manner and style of painting. Look at the creations of Rembrandt of The Netherlands in the 17th century, or of our own tradition of miniature paintings, just before the Industrial Revolution. There is no mistaking the identity of what has been painted. One might say that the identity has not been reduced to its "elements". A part of this identity was lost during the Impressionistic Movement of the 18th and 19th centuries in Europe. And, today, when man and his science are considered to be more evolved than a 100 years ago, this identity has been reduced to its elements. The emphasis in the abstract painting or sculpture of today is on form, colour, composition, geometry, texture, and symmetry (or a deliberate lack of it); its abstraction and symbolism can be compared to the abstraction and symbolism of modern mathematics. As man has evolved intellectually and socially, and as his science has progressed, his artistic creations have tended towards increasing harmony with nature and have, in the process, become more abstract — more like mathematics which is the abstraction of all sciences. One might even speculate that man's pinnacle of evolution would be reached when all his creations achieve total harmony with Nature. That man indeed has been moving — knowingly or unknowingly — towards this end, is also indicated by comparison of the work of man and the work of Nature, both independently created.

A painting done by the well-known Indian painter, Surya Prakash, hangs in the Method of Science Exhibition at Hyderabad. For this painting, as for many of his other paintings, Prakash used what we throw away — garbage. If you compare this exquisite painting with a scanning electron micrograph of the surface of a leaf, you would find that the similarity is remarkable. One is man-made, the other is made by Nature; both are exquisite works of art.

Let us look at another example. X-rays allow us to perceive structures that cannot be seen even by a microscope. In fact, through X-rays we can locate the relative position of each atom in molecule. Thus, today, using X-rays, one can look at the actual shape of a molecule in three dimensions. Whales contain a substance — a protein — called myoglobin, which is somewhat similar to the protein, haemoglobin, that makes our blood red. If you compare the flow of lines in the architecture of a myoglobin molecule derived from whale, as revealed by X-ray analysis, with one of the paintings (of a women) by Henri Matisse — one of the foremost painters of our times — you would find the resemblance, line by line, angle by angle, striking; indeed, it is not a coincidence. It is perhaps this resemblance that makes Matisse a great painter: what he created finds an echo, a counterpart in Nature, and *he* did it effortlessly and intuitively.

The commonality of methodology

There is yet another link between science and art. Creative ability, both in art and science, uses essentially the same method — in science more explicitly, in art more implicitly. The method is, unfortunately, on account of its historical background, called the scientific method. It should, in fact, be called the method for acquiring knowledge, or the method for discovering truth or beauty or the method for finding reliable answers to questions — answers that would stand the test of time. Let us now look at how the scientific method applies to various art forms, including literature.

What is the *modus operandi* for a creative scientist? He must have a question to answer. A question in science, arises out of careful observation or careful analysis of existing information or knowledge.

How does a scientist go about answering a question? He makes various hypotheses. A hypothesis is a possible answer which is testable. In making the hypothesis he brings to bear all his past experience and knowledge.

To test it, he does experiments. If the results of the experiments do not support the hypothesis, he changes the hypothesis. And this process goes on till enough experiments have been done to prove a hypothesis. He would have then arrived at the answer. If the answer is of general applicability, he would have discovered a new law or theory, and could become famous!

An expression of the creative urge in science thus begins with a question something that you do not understand, something that worries you because you do not know the answer. Indeed, it is this "worry" that inspires a scientist to put in the effort he does towards finding an answer.

In the case of a painter, poet or sculptor, too, the creative urge begins with the recognition that there is something unknown and un-understood. Intense observation of, and intimate interaction with the environment, is a hall-mark of the creative artist as of the creative scientist.

The creative urge leads the artist to a new perception of the environment. He finds there is a problem unsolved, a problem unanswered so far, something he does not understand. He feel intensely worried, concerned and disturbed, for nothing moves a creative individual more than something unknown, unexplained or un-understood. "Not everything is right with the world. There is something to worry about, something disturbing or alarming". It is in this process of evaluation and understanding, and of finding an answer, that he brings to bear all his knowledge and experience just as a scientist does. Sometimes, gradually and sometimes, suddenly the creative image builds up in his mind during the above mental exercise — just like a hypothesis in the case of a scientist!

No one — a scientist or an artist — creates just because one is moved. Many may find a beautiful object — a flower, a woman, a natural scenery — a moving experience. But only a few — the exceptional ones — would be "inspired" to create something new following such an experience, for only they would be moved *enough* to be disturbed and thus to think about the "wonder" behind their experience.

If beauty not only moves you but generates in you a sense of wonder, it might lead you to create, like Wordsworth or Cezanne or Amrita Shergil; for, in the sense of wonder, there is a built-in question. Thus, the question and hypothesis (or theory), the first two steps, that a scientist uses, have their counterparts in the creative process of artists — be they poets or painters.

What then? The scientist tests his hypothesis or theory by experiment or, when experiment is not possible, chooses from the several possible hypotheses by determining the compatibility of each with existing knowledge, using logical reasoning. What about the artist? Indeed the creative work of the artist — the book or the painting in the case of a writer or a painter — is his experiment. Whether or not he has succeeded in finding an answer, is what the viewer or the reader would determine.

The answer to questions that a painter or a writer attempts to answer are generally far more complicated than the answers to the questions that scientists ask. Therefore, there may not be clear-cut answers always available to a question through a work of art — a poem, a novel. Yet, if they are products of high level of creativity, they always take man forward, by making the person who understands them more knowledgeable. And is not one of the main purposes of obtaining answers to questions, precisely this — that is, to make you more knowledgeable so that you may, in the future, be able to handle questions that may not even have been asked till that time?

A concern, an awareness of the environment around you, and the recognition of the fact that there are unanswered questions, is a prime motivation for all creative activity. The question may pertain to natural phenomena, as in the case of natural and physical scientists or to social phenomena, as in the case of writers. Indeed, painters and writers, like many others engaged in creative work, derive their inspiration from the environment around them. Therefore, it would not be wrong to conclude that a person who has no "worry" or "problem" which transcends personal interest (this does not refer to trivial personal problems), can never be creative.

In other words, the creative urge, whether in science or in art, arises from the same sort of inspiration and, eventually, through the use of the same method, leads to addition to our knowledge and experience, the interplay of which leads to human progress.

Conclusion

Nature is an embodiment of scientific laws. The order it creates through the application of these laws is beauty; therefore, science and beauty are inseparable. Neither scientists nor artists of any description would create if they did not have the ability to perceive this order. A good scientist has an intuitive eye for beauty and is excited by it, like a good painter, a good writer, or, for that matter, any individual who engages in creative activity. Given a choice between theories, a scientist intuitively chooses the one that appeals the most to his sense of beauty. J.W.N. Sullivan, the author of biographies of both Newton and Beethoven, wrote (in the *Athenaeum* for May 1919 as quoted in an article in *Physics Today* 1979).

"Since the primary object of a scientific theory is to express the harmonies which are found to exist in nature, we see at once that these theories must have an aesthetic value. The measure of the success of a scientific theory is, in fact, a measure of its aesthetic value, since it is a measure of the extent to which it has introduced harmony in what was before chaos. It is in its aesthetic value that the justification of the scientific theory is to be found, and with it the justification of the scientific method".

Indeed, by choosing a theory that has an intrinsic beauty, the scientist increases the chances of his success, as history has shown many times.

In an article in *Physics Today* (July 1979), the Nobel Prize-Winning astrophysicist, S.Chandrasekhar cites the case of Hermann Weyl's "gauge theory of gravitation", which he had worked out in his *Raum-Zeit-Materie*. Apparently, Weyl became convinced that this theory was not true as a theory of gravitation, but still it was so beautiful that he did not wish to abandon it, and so he kept it alive for the sake of its beauty. But much later, it did turn out that Weyl's instinct was right after all, when the formation of gauge invariance was incorporated into quantum electro-dynamics. Another example Chandrasekhar gives is of Weyl's two-component relativistic wave equation of the neutrino. Weyl discovered this equation and the physicists ignored it for some thirty years because it violated parity invariance. And again, it turned out that Weyl's instincts were right. Thus, creativity in science demands the ability to perceive beauty, as much as creativity in the sphere of any other endeavour. An elegant proof of a theorem is far more satisfying than a cumbersome proof.

Indeed, practitioners of both science and art are seekers of beauty and all seekers of beauty have a tryst with truth. Contemporary art represents progress in art — as contemporary science represents progress in science. There is no doubt that man's eternal quest for truth would lead him to open new vistas in science and art, as it would lead scientists and artists to open new vistas of beauty. Progress will continue to occur as long as we can perceive and recognize beauty through our five senses, and then understand what we have perceived, through the use of science, or of art, or of both — for both are becoming increasingly inter-dependent, in terms of, say, materials, ideas and ideals. Science and art are just two exchangeable currencies; use whichever you will : that is not important. What *is* important is to know what you wish to buy, and what its real worth is !

As Keats has said :

> Beauty is truth,
>   truth beauty — that is all
> you know on earth,
>   and all ye need to know.

And as Einstein said to Heisenberg :

"If nature leads us to mathematical forms of great simplicity and beauty — by forms, I am referring to coherent systems of hypothesis, axioms, etc., — to forms that no one has previously encountered, we cannot help thinking that they are "true", that they reveal a genuine feature of nature... You must have felt this too; the almost frightening simplicity and wholeness of the relationships which nature suddenly spreads out before us and for which none of us was in the least prepared".

The terms symmetry, beauty, science and truth may have had independent origins, but they are so closely related — the overlap is so enormous — that it is difficult to perceive where one ends and the other begins. They pervade each other in a way that their identity would seem to be lost. To recognize this, is an enriching and rewarding experience.

Limitations of the discussion

1. We have deliberately not touched upon music. There are some very interesting new findings that might give us an insight into the origins of music. One of these is a recent report in which the genetic code has been shown to be capable of being set to music. This implies that the patterns that one finds in music are not confined only to music.
2. We have not discussed the role of the context in which beauty is perceived. Our perception of beauty is no doubt modulated very considerably by the context — external or internal — in which the object is perceived.
3. We have also not discussed the sociology of beauty or aesthetics. The response to beauty is so deep-rooted that it often leads to a bias which is extremely difficult to overcome in our everyday life.

The Symphony of Symmetries

K. Babu Joseph

The word symmetry has a plethora of meanings: it denotes the quality of harmony or agreement as well as the existence of right relations between parts. Symmetry also signifies the beauty resulting from proper kinship. In science, the term is often used in the sense of an operation that leaves an object or a process invariant. Thus it has both static and dynamic connotations. One of the perennial problems of philosophy is beauty, and as Anatole France put it, we shall never know why a thing is beautiful[1]. Yet, at the risk of being branded a reductionist, one might be inclined to analyze a thing of beauty in terms of its symmetries and discover joy for ever. According to Aristotle, beauty is symmetry, proportion and an organic order of parts in a unified whole[2]. Nature reveals her splendour through a rich collection of symmetries in relation to structure, organisation and hierarchy. One usually associates symmetry with order, form, regularity, rhythm and harmony. These qualities may be found not only in individual objects and phenomena but also in society and its evolutionary processes.

Symmetries are important in art, music, religion, literature, science and philosophy. The existence of symmetry in natural objects and natural phenomena was noticed in early human civilization. The primitive man's tendency to associate a god with every natural phenomenon was probably one of the earliest applications of the symmetry principle, for it is based on a conjectured symmetry between cause and effect, that is, to every effect there must be a cause. The symmetry between right and left, circular symmetry and repetitive spatial symmetry found expression in ancient handicrafts and architectural motifs. The melodies and harmonies of music, the patterns of events unfolding in countless legends and myths, the rhythm of day and night, the orderly transformations of the moon, the cyclic appearance of seasons, and several other phenomena, have from times immemorial provided grist to the symmetry mill. One can even discern an unconscious application of the symmetry motif in ancient constructions of geocentric and heliocentric models of the planetary system. Both classes of theories presented highly symmetric configurations of objects in the sky. It is no exaggeration to suggest that in certain societies symmetry considerations influenced the emergence of the concept of God as the sole creator and arbiter of the world.

Coming to modern times, symmetry has played a vital role in the development of chemistry and physics. Of late, symmetry has been recognised as an important aspect of molecular biology as well. The formal study of symmetries is called group theory which constitutes an outstanding achievement of modern mathematics. Since it is not possible to deal with such a *pot-pourri* of topics here, we shall confine ourselves to some recent trends in the study of symmetries. Using the language of contemporary physics, we shall introduce a few rudimentary types of symmetry and briefly discuss the concept of spontaneous symmetry breaking and its applications. The connection between symmetry and unity will be highlighted. We shall also attempt a concise application of the symmetry paradigm to creative endeavour such as art, literature and spirituality.

Symmetries in physics

Being a fundamental discipline, physics provides a convenient framework for launching a sensible discussion on symmetries. If an object or a process remains unchanged under a specified operation such as translation, rotations, reflection, magnification, etc., the particular operation is called a symmetry operation and constitutes a symmetry or invariance of the object or process. A self-contained set of symmetry operations forms a symmetry group. This means that any two operations of this set can be combined to yield a third operation belonging to the set, and the 'inverse' of each operation in the set is also contained in it. Empty space possesses all the above mentioned symmetries. It permits arbitrary translations, rotations, reflections and magnifications. This implies that a lone object situated in empty space can be translated by an arbitrary amount, rotated through an arbitrary angle about any axis, reflected in any plane or magnified to any scale, without altering its properties. These symmetries are generally valid for macroscopic objects insulated from external influence. At the microscopic or sub-atomic level, there are some limitations on the validity of reflection or right-left symmetry and the imagined operation of 'going backward' in time. Certain forms of radioactivity caused by the *weak force* do not respect these invariances. The breakdown of right-left symmetry or the development of 'handedness' is of some importance in biology and medicine too. A drug such as adrenaline which is used to contract blood vessels is effective in one form and ineffective in its mirror image.

Einstein's theory of relativity introduced a symmetry between space and time in such a way that whatever statements apply to space are true for time also. This symmetry called relativistic symmetry or Lorentz symmetry, which seeks unity between space and time concepts, is transparent only at speeds approaching that of light. An astounding consequence of Lorentz symmetry is the famous mass-energy equivalence. Theories built on the relativity principle such as the theory of elementary particles, are known as relativistic theories. In a relativistic theory, space-time appears as a single unified entity possessing translation, rotation and other similar invariances. A theory formulated in the space-time scenario is highly non-trivial and differs drastically from one set on a split space and time scenario. Based on permutation symmetry, quantum physics envisages a division of identical particles into distinct classes called bosons and fermions. Although there is so far little empirical support, considerable amount of theorizing has gone into relating bosons and fermions under a unified scheme called super-symmetry.

Symmetries can be broadly classified into external and internal types. An external symmetry corresponds to an invariance under an operation — for example, translation — that is thought to be carried out outside the object. Space and time or space-time (in the sense of relativity) symmetries are external ones. Internal symmetries have so far remained an exclusive frontier of high energy physics. An internal symmetry requires invariance under transformations in an internal abstract space. Properties of elementary particles, such as isospin, strangeness, hypercharge, charm and beauty, are examples of those derived from symmetries reigning in various internal spaces. These have been understood to be only approximate symmetries and called flavour symmetries. It is remarkable that several new particles have been predicted and identified with the help of flavour symmetries. However, another type of internal symmetry called gauge symmetry is of paramount importance in fundamental physics today. The existence of the electro-magnetic field, which is perceived as light, is believed to be a consequence of a simple gauge symmetry that governs the behaviour of charged particles. A mathematically more complicated form of gauge symmetry has been postulated to be responsible for the strong forces that bind atomic nuclei.

Though exact under ordinary circumstances, a gauge symmetry may be broken by a process called spontaneous symmetry breaking (SSB). Indeed any symmetry may be broken by this process. For illustration, consider a ferromagnet, say a piece of iron, in which the atomic magnets point in the same direction. Such a configuration violates symmetry, because it selects a definite direction in space. On heating this specimen, the atomic magnets begin to tumble and at a critical temperature assume random orientations. When this happens, rotation symmetry is restored spontaneously. Under the reverse process of cooling below the critical temperature, rotation symmetry is spontaneously broken. Certain gauge symmetries are believed to have been spontaneously violated as the universe cooled from an initial state of very high temperature. These ideas underlie much of the recent pursuit of unified theories of the four fundamental forces, namely, gravity, weak force, electro-magnetism and strong force. In spite of the fact that these forces are observed to be distinct in the present universe, they must have remained unified, reminiscent of Aristotle's conception of beauty, at the beginning of cosmic evolution. A great deal of available empirical evidence shows that at an earlier epoch at least two of these forces, namely, the weak and electro-magnetic forces, constituted a single unified force when the temperatures were pretty high everywhere. The unified electro-weak theory, as it is called, makes use of a complicated scheme of gauge symmetry and its spontaneous breakdown at a sufficiently low temperature.

During the past two decades, considerable effort has been spent towards constructing grand unified theories that seek the unification of all forces. Contrary to the first flush of euphoria, none of these schemes has met the stringent criteria of experimentation. The latest model in this arsenal is super-string theory which, too, has failed to give any verifiable predictions. Despite a woeful lack of empirical confirmation, these models possess an enormous tower of internal symmetries. The unification scenario predicts or rather retrodicts that the initial state of the universe enjoyed a supreme level of symmetry which subsequently broke down spontaneously in different stages of evolution of the universe. The message that these theories carry is that the higher the degree of symmetry, the greater would be the degree of unity or simplicity; complexity, on the other hand, is a natural symptom of low symmetry.

Exciting developments are currently taking place in the study of open systems which are in constant communication with the environment. Open systems include the whole living world as well as most non-living things. The main thrust of this research consists in understanding the spontaneous generation of spatial and temporal patterns through some mechanism of symmetry breaking. A flowing liquid heated from below, for example, is seen to develop at a sufficiently high temperature, a variety of spatial patterns such as rolls, squares, hexagons and so forth. Chemical oscillations occurring in some reacting chemical systems illustrate the formation of temporal order. The growth of tissue patterns, colours of butterfly wings, stripes on animal furs and countless other phenomena can be interpreted in terms of the symmetry breaking metaphor.

The symmetry group of a specific spatial pattern is definitely *smaller* than that of empty space, to which reference was made in the opening paragraph of this section. A smaller group contains fewer operations than a larger one. For a circular pattern, for example, any diameter or a vertical axis passing through its centre is a symmetry element — namely, one which defines a symmetry operation; in this case, rotation — but an axis passing through its circumference is not. Contrast this with the case of empty space which permits rotations about any axis. It readily follows that the symmetry group of a circle is smaller than that of empty space. In technical parlance — the symmetry group of a circle is said to be a subgroup of that of empty space. Patterns may arise in open systems when the symmetry group is chopped down to a smaller size, and the contention of protagonists of SSB is that this can happen spontaneously through interaction with the surroundings.

Self-similar patterns have recently become the focal point of much scientific attention. Such patterns are called *fractals*. Clouds, snow flakes, foliage, river edges, mountain ranges, etc, are fractals. A fractal exhibits the same structure at all scales of observation. Not every fractal is a static pattern, for fractal-like patterns may be associated with specific dynamic phenomena. The recent discovery of self-similar jet production in high energy collisions of particles provides a case in point. What is interesting about a fractal is that, to characterise it, it may be necessary to employ some notion of random symmetry. Future studies alone will be able to solve the problem.

The existence of a particular symmetry sometimes gives rise to a law such as the conservation of energy which is the result of invariance under displacements in time. When a symmetry is broken, the associated law is also violated. A dissipative system does not conserve its energy because of its constant interaction with the environment, which involves energy exchanges. For such systems time translation is not an allowed symmetry.

Symmetries in human endeavour

As already mentioned, symmetry affords a convenient phrase for discussing order of all kinds, namely, order in nature and order in human creations. This by no means suggests that straight-forward applications of ideas from physics to humanities and social sciences are always true. The element of randomness inherent in human activities gives the customary explanation for divergences from physics-inspired projections and extrapolations. However, it would be quite relevant to notice that determinism is at a low ebb today even in much of physics. Large areas of classical physics, let alone the whole of quantum physics, are now known to lead to indeterminism. In several classical systems which were once thought to be deterministic, chaos is often the rule rather than the exception. Most dissipative systems found in nature are of this category. Keeping in mind such limitations of predictability in the universe, we turn to a few applications of the symmetry principle to non-physical systems.

Art is a domain wherein the concept of spatial symmetry can be readily invoked. For a symmetry-analyst, a painting or a piece of sculpture is just a display of a subgroup, of the symmetry group allowed for empty space. The smaller the subgroup the greater is the apparent complexity of the work. There is an interaction between the artist and the raw material used for the creation of the work that culminates in the emergence of a sub-order which has remained unnoticed in space before. The new order or symmetry group presented by the work is not all that makes it valuable. As everyone would agree, the merit of the work is mainly determined by the emotions that it evokes in the artist as well as other viewers. These reactions may be thought of as patterns that live in some inner abstract space. Since the mood of each viewer of the work is not constant, these patterns develop dynamically. Bohm and Peats have proposed the term *explicate order* for that which is manifested and the term implicate order for that which remains hidden[3]. Geometrical symmetry represents explicate order and inner space symmetry, implicate order. Aesthetic judgement is a process that takes place in the inner space of emotional responses. It must be emphasized that the inner dimensions of a work of art exist in the minds of the audience and not in the object. An object will qualify as an artistic creation only through the conjunction of external symmetry manifested as geometrical symmetry and internal symmetry expressed through emotional responses. The interpretation of art as a material object forms the basis of our whole discussion. Popper defines world 1, as that of material objects, world 2 as that of our subjective experiences and world 3 as that of our arguments, interpretations and theories[4]. Geometrical symmetry is a property of world 1, while the internal symmetry of responses operates in a union of worlds 2 and 3.

A strategy similar to the one above is applicable to poetry, drama and literature in general, not to speak of various forms of visual arts. In each case, external and internal symmetries exist in the sense already discussed. An outstanding problem is how to characterize these internal symmetries. Since this is a highly technical question, we shall content ourselves by saying that the internal symmetry resides in patterns of emotional response. Some of these symmetries are truly complex, partly because even though an object or a literary piece may be physically stationary, it can stimulate a sequence of dynamic reactions in the form of widely fluctuating patterns. Consider one of the well-known dances of *Siva* called the *Tandava*, representations of which are found, for example, in the *Ellora* and *Elephanta* caves. The dance brings out *Siva's tamasic* aspect[5]. An idol in this pose might stir a volley of emotions: fear, revulsion and sheer helplessness against nature's fury. These form a pattern or a sequence of patterns involving other objects or situations which are capable of arousing kindred feelings. With the passage of time, the viewer might normally realise the utter unreality of most of these patterns and shake himself or herself free from their grip as he or she should, because inner space patterns are dynamic and as a result they grow, decay and disappear. Such changes often take place autonomously, implying a sequence of SSBs and restorations. Growth and death, biological or otherwise, are phenomena linked to SSB and restoration, respectively.

Symmetry ensures some kind of stability as in balanced economy or peaceful polity. The laws of a country envisage a healthy society. Any activity permissible under law is a symmetry operation for individuals and society at large. War seeks to destroy not only nations but also regional and, sometimes, international power equations, and are symmetry breaking phenomena. Structure, organisation and hierarchy are manifestations of symmetry of one form or other. Of these, structure is the most fundamental thing; organisation and hierarchy follow it. According to Haken, one of the founders of the discipline called synergetics, structure can arise in a system composed of a large number of subsystems which interact with one another coherently[6]. Quite often, concerted action can result in qualities not present in the individual subsystem, signalling the formation of a new symmetry, which, on the scale of symmetries introduced earlier, may be of low order. Self- organisation is the favourite term for this phenomenon for scientists such as Haken[6] and Prigogine[7]. A nation after passing through the convulsions of a revolution may settle down to a new social order which could be a specific pattern of symmetries and balances.

Unity, peace and bliss

In the physical universe, symmetry implies unity as revealed by the fundamental forces. Diversity or structure arises through its spontaneous breakdown bringing in complexity and, in the sense of information theory, information. Conversely, a unified state possesses little or no information. Thanks to the close kinship between unity and symmetry, one need only search for one of these; the other will naturally follow. Here is a quote from *Katha Upanishad*:

"The sense-objects are higher than the senses, and the mind is higher than the sense-objects; but the intellect is higher than the mind and the Great Soul is higher than the intellect".

Were one to interpret this within a symmetry framework, one would have to assign the highest form of internal symmetry to the Great Soul. The various other structures referred to here possess progressively lower grades of symmetry. What is remarkable here is the indian psyche's vision of a grand hierarchy with the Great Soul sitting at the zenith and the sense-objects at the nadir.

To a seeker of enlightenment, the discovery of the Great soul is not enough. He or she must imbibe the supreme level of symmetry implied by the cryptic statement : *Aham Brahmasi* — I am the Brahman (the Absolute). To the quantum-mechanically minded person, this may be transcribed into : 'the observer and the observed are one'. Both become unified in the highest state of symmetry.

That unity leads to paucity of information or absence of quality seems to have been appreciated by ancient Indians. For the concepts of *Nirguna Brahman* and *Saguna Brahman* merit an examination within this frame. *Nirguna Brahman* is the ultimate; even the name should not be uttered! *Saguna Brahman* belongs to a lower stratum, because it possesses qualities. The degradation of the *Nirguna Brahman* to *Saguna Brahman* may be regarded as a spontaneous breakdown of symmetry. One might expect that this philosophical idea contains the seed of a great vision of cosmological evolution.

Meditation is the traditional route towards achieving self- realisation, that is, unity of the observer and the observed. It involves the sealing off of all doors of perception and indulgence in concentration. In the issuing final state — successful meditators are said to slip into a trance, that is, a thought-vacuum ¤ no qualities, no information, but only symmetry, unity, peace, and bliss.

References

1. W. Durant. *The Pleasures of Philosophy*, New York : Simon and Schuster, 1929 and 1953.
2. *Ibid.*
3. D. Bohm and F.D.Peat, *Science, Order and Creativity*, Toronto : Bantam Books, 1987.
4. K. Popper, *Unended Quest*, Glasgow : Fontana Paperbacks, 1986.
5. A.K. Coomaraswamy, *The Dance of Siva*, New York : Dover, 1985.
6. H.Haken, *Complex Systems* — Operational Approaches, Berlin : Springer-Verlag, 1985.
7. G. Nicolis and I. Prigogine, *Self-organisation, in Non-equilibrium Systems*, New York : John-Wiley and Sons, 1977.

Education through Astronomy

Jayant V. Narlikar

To the layman, astronomy is often mistaken for astrology and vice versa. He often brings all his superstitions to bear whenever the science of astronomy is being discussed or described. To the scientists in general, non-familiarity with astronomy makes the subject appear just a collection of speculative ideas.

The purpose of this paper is to highlight the vital contribution astronomy can make as a regular part of our educational curriculum and also towards the enlightenment of public at large. This is because astronomy possesses certain unique features that no other science can claim to have.

Let me illustrate this with an analogy. A person confined to his native village has a limited outlook and experience. If he moves on to the district town, he receives a quantum jump in his overall perception. Travel broadens the mind — so they say! If he travels to a metropolitan city, then through the state and the nation, then through the continent and eventually jets across different parts of the world, these quantum jumps of knowledge continue in ascending order. So it is with astronomy *vis-a-vis* other sciences.

The typical laboratory scientist confines (and is constrained to confine) himself to experiments, observations and inferences here on the Earth. While he can claim to be in control of whatever experiment he performs, the terrestrial confinement necessarily limits his vision of nature. Astronomy takes him away from his native Earth out into space. The farther he moves, the wider is his perception of the universe. The quantum jumps of knowledge occur as he encounters the hierarchical structures in his journey through the universe: the planets, the stars, the galaxies, the clusters and superclusters of galaxies .... and so on.

The remarkable feature of astronomy is that even though man sits here and observes these celestial phenomena from afar, he is able to make some sense of what he sees on the basis of the science he knows. "The most incomprehensible thing about the universe is that it is comprehensible", said Albert Einstein reflecting his astonishment at the apparent universality of science.

This universality has two consequences. First, through a proper application of science, one begins to understand something that may have appeared mysterious, awe inspiring and even frightening to start with. Many superstitions arising from a lack of proper understanding of the cosmic events are thus washed away. Secondly, the scientist himself gains confidence in his laws of science which, after all, have been derived from empirical observations of nature in the first place. The cosmos provides him with a far bigger and far more versatile laboratory than the Earth could ever do. And so, through astronomy he gains a better understanding of science itself.

It is surprising therefore that these two benefits have not been properly assimilated in our educational system, either the formal one of schools and colleges or the extra-mural one relating to the general public. By integrating astronomy into other educational inputs we should be able to take advantage of what the subject has to offer. Indeed, I may go further and say, we must do so!

In the rest of this paper I will present illustrations of specific questions through which astronomy can generate important educational inputs. I will consider three different types of questions.

1

Motion through Space

i. For us, the inhabitants of the Earth, the planetary surface appears fixed. However, it is not fixed within the cosmic stellar background. The Earth spins around its polar axis from west to east as a result of which we see the Sun rise in the east and set in the west. *But can we ever see the Sun rise in the west?* This question would immediately draw an answer 'NO'. The correct answer in fact is 'YES'. The trick lies in achieving motion around the Earth that would overcompensate for its spin. This is easily achieved at high altitudes by travelling in a jet plane from east to west. I had this experience while travelling from London to Chicago, when around dusk I noticed that the Sun, which had earlier set in the west, rose again and went up above the horizon for a few minutes.

ii. *Why does the Moon appear to follow us as we move on the roads at night*? Any stationary object on the Earth in our neighbourhood changes its direction:

but not so the Moon, because it is very very far away compared to the terrestrial distances we travel in our journeys. This has a historic parallel that needs to be pointed out. The Greek thinker Aristarchus of Samos (around 310-230 B.C.) had the intuition to argue that the Earth is not fixed and that it moves through space against the stellar background. To verify his contention, Hipparchus (130 B.C.) had proposed observations of directions of stars. If the Earth moves round the Sun in a circular orbit (so assumed for simplification), then it is farthest from its contemporary position six months later. Thus, the star's direction would be changed when observed now and six months later. No such change was actually found. This was not because Aristarchus was wrong in principle. He had underestimated the stellar distances and overestimated the effect. The observing techniques of those times were incapable of measuring the very small changes of direction which could only be confirmed by the parallex measurements of the nineteenth century.

iii. The shadow effect that leads to solar and lunar eclipses is described in the school texts. What need to be described also are the relative angular sizes of the Sun and the Moon and the fact of their being nearly equal. *What role does this fact play in producing a total solar eclipse?*

More interesting is to imagine the relative appearances of other planets and their satellites when viewing one another. *How large would the planet Jupiter look when viewed from its nearest satellite Io*? The angle subtended is 38 times that subtended by the Moon at the Earth. In other words,

Jupiter would look a disc of thirty eight times the size of the lunar disc.

The Moon rises and sets over the Earth. *How does the Earth appear from the Moon*? It looks as pretty as the Moon does from here. *But will it appear to rise or set*? The answer is not what you would normally expect.

iv. Comets have inspired the astrologers to make dire predictions about disasters on the Earth. As Carl Sagan once put it, the only correct prediction about a comet was first made by an astronomer and not an astrologer! It was Edmund Halley who successfully calculated the passage of the comet close to the Earth, the comet that has now been named after him.

Indeed, Halley's calculation of the cometary trajectory was a direct application of a predictive scientific theory (Newton's laws of motion and gravitation) to observed data. The scientific nature of astronomy can thus be illustrated with this example.

2

Physics of the local environment

i. By local environment, I mean the immediate surroundings of the Earth. A question which excited no less a person than Sir C.V. Raman to think about the answer is: *why is the sky blue*? The role of scattering and absorption of light by the particles of the atmosphere can be explained to the questioner in the course of giving the answer to this question.

A related question is: *how would the sky look from above the atmosphere*? If we consider light from the Sun, it would reach us straight, as there are hardly any scatterers in space. Thus, the Sun would appear as a bright ball, while the rest of the sky is dark! This is why with a space telescope we can do astronomy twenty four hours a day.

It is the atmosphere again that leads to the twinkling of stars. *What is the visual difference between the appearance of a star and a nearby planet?*

ii. Light is actually an electromagnetic wave and the appearance of light — its colour if it is visible — depends on wavelength. It is important to describe (even in a qualitative manner) the different types of astronomies at different wave bands. The radio astronomy, X-ray astronomy, the ultraviolet and infrared astronomies and gamma ray astronomy are all possible, thanks to the technological developments since World War II. *How have these astronomies added new dimensions to human understanding of the universe?*

A major input has, of course, been possible due to space technology. The atmosphere absorbs most electromagnetic waves, except those in the visible and radio bands. So to "do" all the other astronomies one needs to site their respective telescopes above the atmosphere, in satellites.

Once you are above the atmosphere, it pays to 'look down' on the Earth too. Why? Because, from a distance you can get a global perspective of the planet that no view from a short height can give. Here it is useful to give the analogy of how a view of a terrain from the top of a tall tower is more comprehensive than the information gained by walking on it. In short, we do not get a proper perspective from too short a distance : the remote sensing program has a valuable role to play.

3

The relevance of astronomy to our existence

i. That astronomy is not a subject reserved for the starry eyed speculators only can be brought home through its applications to human life. To begin with, life on Earth would not be possible without the primary source of energy, the Sun. The Sun is at the 'right' distance from us for this purpose; neither too far, nor too near.

But how does the Sun generate energy? Here it is worth giving an account of how scientists solved the problem, first with some unsuccessful ideas. The final successful solution — nuclear fusion in a controlled form — was offered by astronomers and not laboratory physicists. Even today, it has not been possible to achieve nuclear energy generation through controlled fusion in a man-made laboratory. *Where and how does astronomy score in this game*?

ii. Take a stainless steel spoon. *How was it created*? In a factory, out of steel, Where did the steel come from? From a steel plant. How did it get there? From raw ore in the mines of the Earth. *In what way did the Earth acquire it*? Probably it was present in the material from which the Solar System formed. *But how did it get into that material*? Probably dumped from the debris of nearby exploding star — the supernova.

This brings us to the question of nucleosynthesis in stars. The deep interior of a star serves as a fusion reactor. At temperatures of several million degrees, lighter nuclei can be fused to make bigger ones. The iron that forms part of your spoon was made in some such star which has already exploded and dumped its inner core material out into the interstellar space. The story of how supernovae are observed itself is of interest, starting from the discovery of the Crab Supernova of 1054 A.D. to SN 1987A, the supernova seen only three years ago. How scientists could measure the neutrino fallout from a supernova with their newly set up detectors is a success story of astronomy as a scientific discipline. The neutrino emission is predicted by theory to happen before the star explodes. This is precisely what was found in the case of SN 1987A.

What would happen if another supernova explodes in our neighborhood, say within 30 light years? The fallout would also include high energy cosmic rays whose impact on our atmosphere can destroy the protective ozone layer. Thus, a supernova might one day destroy life on Earth just as another was probably responsible for the origin of the Solar System.

iii. The physicists need to be reminded that one of their basic interactions was detected and tested through astronomical observations. The law of gravity was inspired and tested by the motions of planets and satellites, and not by any laboratory experiment.

It is the law of gravity that forms the basis of all motions of the spacecrafts launched by man in the Solar System. The accurate trajectories of lunar crafts, the interplanetary projects like the Voyager and the geostationary satellites that play such a key role in modern communications are computed with full faith in the law of gravity.

4

Concluding remarks

These are some instances which demonstrate the need to integrate astronomy with education, either formal or extra-curricular. It is essential to provide astronomical inputs early in human life if only to give man a better understanding of the universe he lives in and a realistic perspective of his place in it.

The methods of bringing in astronomy need not be confined to books alone. Although there is great shortage of suitable introductory literature, Video programs can bring the rich visual material currently available in astronomy to the learner much more effectively than the written word can. Such programmes can succeed if they link astronomy with human existence and encourage the innate curiosity. It is through such educational experiments that we may successfully combat superstitions. For, in the last analysis, a superstition can be destroyed not just by demonstrating that it is based on false premises but also giving in its place the correctly reasoned scientific picture.

Inter-University Centres and Undergraduate Science Education

V.G. Bhide

Science has, over the years, come to occupy centre stage in the thinking and living of man. The uppermost thought that comes naturally and spontaneously to the mind of a conscious being, and which returns after every banishment, is the one concerning the dominating role of science in shaping man's view of the universe, his sojourn on this planet and his ultimate destiny. For good or for bad, science has permeated deep in all the sectors of human activity and hence it is not only opportune but also essential that we consider in all its aspects the implications of science in society of which education in science and science in education are the most important components.

The world is a construct of sensations, perceptions and memory of the interaction between living beings and the interaction between matter and energy within the confines of space and time. Our knowledge of the universe concerns the knowledge of the society in which we live and the world of things we see and deal with.

The former constitutes the whole gamut of social sciences, whereas the latter are pursued by natural sciences. In fact, both science and social institutions have profoundly affected man's life on this planet. Science and society have interacted and will continue to interact in a variety of ways and the tendency to emphasize one or the other has been the cause of much of the recent controversy as to their mutual relationship. The easiest to analyse and perhaps the most obvious is the impact of science on society, although the impact of society on science has been equally important and profound. The impact of science on society takes many forms. The most profound of these is the change in man's view of the universe and the nature of things brought about by science. However, the most visible impact of science on society has been through the applications of science to achieve societal objectives. This application of science for the production of goods and services which has made man's life on this planet happier, healthier and easy is referred to as technology. However, one must realize that science of today is the technology of the morrow. The most important and yet invisible and often neglected role of science is the one concerning its cultural aspects, the way to deal with problems, the methods of acquiring knowledge, the objectivity in dealing with situations, etc. In a sense, science has two aspects, the product of science and the process of science — the method and the content. It is both these aspects which should be adequately reflected in the process of education.

Science and technological basis of modern world, more than anything else, has made education, as never before, the most important element in the life and progress of the nation. Economic and social development, welfare and security are closely dependent on the extent and the quality of education. It is not mere rhetoric when I say that the destiny of India is being shaped in its classrooms and its laboratories. It is on the quality and the nature of the persons coming out of our schools and colleges that the success of our efforts in national reconstructions and social transformation will depend. Pandit Jawaharlal Nehru, once said, "It is science alone that could solve the problems of hunger and poverty, of insanitation and illiteracy, of superstition and deadening customs, of vast resources running to waste of a rich country inhabitated by starving millions". Homi Bhabha, the father of Indian Atomic Energy Programme in his address to the International Council of Scientific Unions (ICSU) in Bombay said, "what the developed countries have and the underdeveloped countries lack are modern science and an economy based on modern science and technology. The problem of the developing and the under-developed countries is, therefore, the problem of establishing modern science in them and transforming economy to one based on modern Science and Technology. An important question we must consider is whether it is possible to transform the economy to the one based on modern technology, developed elsewhere without at the same time establishing modern science in the country as a live and a vital force. If the answer to this important question is in the negative, and I believe our experience will show that it is then the problem of establishing science as a live and a vital force in the society is an inseparable part of transforming an industrially and agriculturally underdeveloped country to a developed country". While these statements are as relevant today as they were when they were first made, problems of national development and the nature of science itself have assumed greater dimensions and complexity and call for not only a restatement of our conviction but also pose a challenge for appropriate and effective action.

Modern India is brimming with young people full of vitality, talent, enthusiasm and aspirations. This is our most precious resource. It is a fundamental right of these young people to go through an education system which motivates them in the right direction, trains them and prepares them for the future and help them in unfolding their potential to the utmost. It is our duty to provide them with such a system. The present conditions are completely contrary to this. Our young people come out of the universities as frustrated individuals who are disillusioned with their surroundings, fearful of their future and with no sense of confidence or direction.

The national policy on education approved by the Parliament in 1986 restates the pivotal role of science education in the context of development of scientific temper and economic development. There cannot be two opinions as to the need for establishing highly efficient and stimulating science education programmes so that the growing needs of highly qualified and motivated scientists and technologists can be met effectively. There are various aspects of the total system of science education, namely, in primary science education, science at secondary level, science education at the undergraduate and postgraduate levels and scientific research. Although one can discuss at length, each of these topics, the present article will be confined mainly to science education at the undergraduate level and science as an exploration which happen to be weakest links, and on Inter-University Centres — a new innovation of the University Grants Commission.

Before Independence almost all the scientific research, and whatever little technological development efforts were being made, were all carried out in the universities. Indeed, universities were at the centre of gravity of almost all the academic activity in the country and it provided leaders in almost all the fields of human endeavour. C.V. Raman, M.N. Saha, Birbal Sahni, S.N. Bose, S.S. Bhatnagar, K.S. Krishnan, D.S. Kothari and several others whose work earned name and international recognition worked in the universities. However, after Independence the situation changed drastically. In our zest to create in short time science as a live and vital force, a number of national laboratories and agencies were created outside the university system without any linkages whatsoever with the universities and unfortunately caring precious little for establishing a sound base of science education to build a superstructure on. Being few and relatively new, these national laboratories were funded liberally, siphoning men and resources from the university system with the consequence that the universities have been starved beyond recognition. The disparities in the resources and infrastructure available in the universities and in National Laboratories need to be seen to be believed. The situation was further complicated by the unplanned tremendous expansion of the university system with their outmoded management and administrative practices. All this resulted in the continual decline of standards of education and research. For worse, universities became a hot bed of politics giving rise to student unrest, tensions on campuses, sometimes bordering on chaos. All the excitement, creativity and motivation disappeared from the classrooms and laboratories. It is no wonder that students who are trained in this atmosphere come out of the universities disillusioned, fearful, lacking confidence and sense of direction and purpose. Further, the universities in the process were left untouched by the spectacular developments that have taken place in many fields of science and technology and were isolated from the mainstream of science and technology. I believe that when resources are meagre, as they are in India, it is prudent to plough them in the universities which have an inherent multiplier effect because of the continual inflow of young boys and girls. It is universities and not bureaucratic research laboratories that can provide an atmosphere conducive to enquiry, search for truth and widening the horizons of human knowledge. It is essentially an uninhibited dialogue between young and enquiring minds and an experienced professor that can lead to discovery and innovation. In contrast, as someone has said, a research laboratory created for a certain cause or round an individual loses its significance and relevance after passage of time and tends to bolster its image by laying claims on higher resources for marble floor and fancy equipment in which it is easier to reach international standards than in academic excellence. We must realize that the educational system is like a mighty Ganga, whereas the research laboratories and agencies are like canals wherein there is a directed flow. It is in the interest of these National Laboratories, industries and hence of the entire country that the mighty Ganga is allowed to flow unpolluted without interference or obstruction. Education and research must go hand in hand, one reinforcing the other, particularly when the time gap between a discovery and its application is shrinking. This separation of education and research as we have tended to do can only be a temporary strategy and not a permanent policy. I strongly feel that we must rebuild our universities and create national facilities on the campuses of the universities, though not necessarily under their administrative control.

The University Grants Commission in its earnest desire to bring universities at the centre stage has launched various programmes, such as University Leadership Programme, Programme of Special Assistance, COSIST Programmes, etc. Some of these programmes, particularly the COSIST Programme, being highly selective have made some impact. The UGC also realized that it should either create or make available to the universities major facilities such as reactors, accelerators, radio telescopes, etc. By their very nature, not only in India but even in advanced countries, such facilities cannot be many and should be optimally used by the entire system. It is this thinking that is at the root of the inter-university centres. It is only recently that the UGC has come to be in a position to establish common facilities as autonomous bodies. The first in the series was the Nuclear Science Centre at the Campus of Jawaharlal Nehru University, New Delhi. This Centre has established a Pelletron as a common facility for the entire university system.The next in line was the Inter-University Centre for Astronomy and Astrophysics (IUCAA). IUCAA is located on the campus of the University of Poona, Pune and aims at establishing internationally competitive centre in astrophysics and astronomy making use of the unique Giant Metre wavelength Radio Telescope facility to be set up by TIFR, near Pune. Excited by the concept of Inter-University system, Prof. Yash Pal, then Chairman, University Grants Commission, approached the Department of Atomic Energy for making available to the university system free and uninhibited use of major research facilities either already established or likely to be established by the Department of Atomic Energy in future. The response of Department of Atomic Energy was very positive. The DAE agreed not only to make these facilities available to the researchers from the universities but also offered to put at their disposal some of its infrastructure. The UGC in turn agreed to establish Inter-University Consortium. It was a memorable day — the 5th July, 1989 and an historic occasion, when a memorandum of understanding was signed between the Chairman of the UGC and the Chiarman of AEC (Atomic Energy Commission) setting out the arrangements between the two agencies and laying down the establishment of the Consortium. The main objective of the Consortium is to provide an institutional framework to enable the maximum utilization of these major research facilities by the university researchers and removing all the barriers — financial, administrative, psychological and others. The Consortium has three centres, one at Indore for Synchrotron Radiation Source, the second at Bombay for Dhruva and the third at Calcutta for Variable Energy Cyclotron.

By their very nature, all the three Inter-University Centres, though differing in details, have a common philosophy and that is to make available to the various universities and institutions of higher learning such major facilities and create conditions for their maximum utilization. All the universities and institutes of higher learning should consider these Inter-University Centres as their field stations and endeavour to formulate exciting research programmes. Through these centres, young students and teachers from the universities will not only come face to face with such major facilities but also with senior scientists from Atomic Energy Establishment, TIFR and other institutes of higher learning. I have no doubt that these centres will facilitate not only what may be called internal transfer of technology but also promote cross-fertilization of ideas, concepts and techniques. Further, the students and teachers from the universities will be exposed to project management, design and fabrication of major facilities and thereby build the necessary confidence and establish the credibility of the university system. Thus, for example, the Inter-University Consortium has taken upon itself to design and fabricate two beam lines on the synchrotron radiation source at Indore. It is not that universities alone will benefit but the Atomic Energy Establishment and other agencies will also benefit perhaps more, because through this arrangement, they will have a continual inflow of young and motivated students. These Inter-University Centres will have a minimal core staff and a vibrant visitor programme, so that at any point of time they will have three or four times the visitors in relation to the core staff.

The signing of the Memorandum of Understanding between Chairman, UGC and Chairman, AEC, is an historic occasion, because it was the first time that the two agencies came forward to share their facilities, expertise and experience in the true spirit of science getting over the normal agency barriers. This concept should be extrapolated to evolve similar arrangements with other agencies, such as CSIR, DRDO, etc. The National Laboratories under CSIR were essentially established to develop technologies for products and processes and to contribute to economic growth. Over the years, these CSIR laboratories have done commendable work. Recently, these laboratories have been asked to concentrate on two or three areas in which they are expected to remain at the cutting edge of technology. By the very nature of the areas that have been identified, the time of obsolescence in these areas is continually shrinking and today ranges between 5 and 7 years. You cannot remain at the cutting edge of technology in these science-based industries unless you are also leaders in the concerned scientific matters. To my mind, the National Laboratories cannot, with their limited resources in men and materials manage to do both. To my mind, a happy marriage between the universitites and National Laboratories could be very fruitful. One can conceive of several models of interaction. Suppose, for example, a National Laboratory has chosen the area of polymers as its major area of operation. The Director and scientists in that laboratory could identify at least five research groups in the universitites which they feel are competent. Then these research groups from the universities and the concerned groups in the National Laboratory could form a composite team and identify the problems of basic nature, applied nature and formulate programmes of action. The university research teams could tackle basic and fundamental problems and the National Laboratory scientists could address themselves to some problems of applied nature, although, of course, one cannot conceive of such hard and fast division of responsibilities. In this process, the university researchers would be brought into the national mainstream, could use the facilities of National Laboratories, work on problems of national relevance, etc. In turn, the National Laboratories would benefit from academic inputs from the universities in the form of ideas, techniques, and more particularly, young students. The National Laboratories should provide facilities to the university researchers to work, whereas the UGC should provide facilities for the university researchers to travel to and from National Laboratories and some research grants for undertaking research on identified problems. I do foresee that in the near future such or similar arrangements could be evolved which would be of immense benefit to both the systems. If this happens, nearly 200-300 research groups in the universities would get involved and it is only to be imagined how beneficial it would be for both. Similarly, some senior scientists form National Laboratories could be visiting professors in the universities.

I would make a brief reference to undergraduate science education and the need to restructure it, so that we not only raise the general level of undergraduate science education but also take deliberate and conscious steps to build peaks of excellence. Since Independence, there has been a phenomenal growth in the number of universities and colleges imparting science education and in the enrolment of students in the science stream. The enrolment in the science stream was about 1,27,168 in 1950-51. This increased to 4,34,925 in 1963-64 and attained a level of 7,25,328 in 1986-87. Thus, there has been a six-fold increase in the enrolment during the last 35 years. Though the actual enrolment has increased steadily, there has been a decline in the percentage of students opting for science. In 1950-51, 32.1 percent students who passed the higher secondary examination enrolled in the science stream. However, this percentage decreased to 19.7 percent in 1986-87. It is not only the decrease in the percentage that should concern us, but the fact that the students offering science today constitute the middle 19.9 percent in contrast to the upper 32 percent in 1950-51. The fact is that the brighter students seem to shy away from science.

The reasons for this decline in percentage and the decrease in the quality of students opting for science are not far to seek. Unlike in 1950-51, the facilities for technical and professional education, such as engineering and medical, have increased considerably siphoning brighter boys and girls to the technical and professional streams, leaving only mediocres to opt for science. The preference for technical stream is because of better employment opportunities for technically trained graduates as compared to those available for a science graduate. In no small measure, the migration of students away from science is due to the dull and drab teaching of science in our schools. The dismal fact is that the persons who do not get admission in professional schools, gravitate to science, not because they are interested or motivated but because they must do something and they do not know what else to do. The brighter ones amongst these also appear in the competitive examinations like, Banking, Indian Administrative Service etc., a different stages of their career. It is this distilled 'left over' that forms the major part of our higher science education enrolment. The above inference is further supported by the follow-up studies of the National Talent Search Awardees. National Science Talent Search Awards were introduced in 1964 to attract bright and motivated students to the science stream. Initially, these nationally selected students were given these awards only if they pursued a career in science. Besides, identifying talent precious little was done to nourish and nurture talent. Unfortunately, however, in subsequent years even this restriction to pursue science was removed. With this change, the input in the science stream suffered qualitatively. The following table gives the total number of awards and those who opted for science.

Year	No. of Awards	Awardees opting for science
1984	375	55
1985	750	121
1986	750	87

The number of talent search awardees pursuing higher science courses is still revealing as per the table below.

	1984	1985	1986	1987	1988
1. B.Sc. Part I	-	63	39	72	69
2. M.Sc. Previous	31	36	13	13	-

Thus, out of the 750 students selected for the talent awards, only 13 went up to the M.Sc. level in 1986 and in 1987. This clearly presents two related problems. A large number of talented students today simply do not opt for science because they do not see it as a rewarding career. At the same time, a large number of unmotivated and uniterested students seek a career in science. To make matters still worse, most of the students, roughly 88 percent, who after having passed the higher secondary examination opt for science education are taught for their undergraduate course in affiliated colleges. Many of these colleges are ill-equipped, over-crowded and poorly staffed. In such an environment, it is almost impossible for a student to experience excitement of doing science and get motivated to take to science as a career. It is really tragic that many of our science teachers, perhaps unwittingly, kill the spirit of scientific enquiry. One often comes across a situation where a particularly harassed and overburdened teacher authoritatively shuts off (even superficial) enquiries from his students. This completely undermines the confidence of the student and subjects him to ridicule. It cannot be overemphasized that creative science progresses only by questioning. Questioning the basics is the heart of creativity. If we prevent the students from asking even the superficial questions, how do we expect them to ask deeper and more fundamental questions? In fact, the questioning spirit of the students should be encouraged.

Similarly, I have come across examples where the students working in a laboratory with a faulty apparatus have gone to the teacher with their not-so-accurate results and have been rebuffed and sent back and then asked to get results as good as of other students. This encourages the students to cheat, to cook the results. Thus, instead of teaching the students the scientific method which would examine the limitations of the apparatus and why the apparatus is giving wrong results and to appreciate these facts, the teacher in fact has forced the students to take a dishonest short-cut completely contrary to the spirit of science. One has not only destroyed the student's ability to do that particular experiment but one has actually damaged his whole understanding of the spirit of the scientific method. It is absolutely essential that teachers who are responsible for teaching science should have a very clear perception of scientific method, spirit of enquiry essential for creative work, honesty and the ability to separate the important questions from the unimportant ones and the ability and the confidence to pursue the question in depth till the answer is found.

Thus, the present system has neither the desired selectivity nor the atmosphere conducive to motivation. The consequences of this are obvious and unmistakable. There has been a continual decline in the standard of our first degree in science. The level attained by our B.Sc. students today is very much inferior to that of a graduate in advanced countries. As a result, science graduates coming out of these colleges lack curiosity, are devoid of creativity and innovativeness, lack confidence in themselves and are almost totally unmotivated. The situation is alarming. It needs no emphasis when I say weaker undergraduate science programme implies weaker postgraduate and research programmes and, indeed, whole of our national effort. It is becoming increasingly difficult in spite of increasing enrolment to find adequate numbers to man our research laboratories, industries, etc.

There are a few more disturbing aspects of our undergraduate science education programme. Undergraduate curricula are rigid even in the best of places. Combinations such as Physics-Chemistry-Mathematics and Chemistry-Botany-Zoology prevent a physics student to enjoy the excitement that exists in life sciences today and in turn it does not allow life sciences students to know and understand the basic physical principles which indeed govern the life processes. Similarly, the undergraduate science education programme depends entirely on chalk-talk concept, doling out uncorrelated information, without bringing out the basic concepts and the basic unity of science. Demonstrations and classroom open-ended experimentation have disappeared. Although science is experiment-based, it is taught as prose or poetry. Further, the examination, laying as it does more stress on memory rather than on testing, (a) capability to analyze situation, (b) reason and logic, (c) comprehension, (d) creativity and model building capacity, (e) communication skill, etc., skews up the entire system.

The task of improving science education programme is stupendous. As mentioned earlier, the yearly enrolment in about 150 colleges is about 2,00,000. It is this large number of science colleges and the number of students enrolled that makes the task of improving science education formidable. I believe that such a stupendous task cannot even be attempted unless one adopts some selectivity and novel approach.

Taking into consideration the genesis of the problem, the magnitude of the problem and the efforts already made, a three-tier approach as indicated below may be considered for adoption. This approach has been guided by the need not only to raise the general level of science education at the undergraduate level but also to deliberately and consciously attempt to develop peaks of excellence.

The first tier programme deals with highly talented students. It should be restricted to the chosen few who demonstrated competence in creative pursuit. On completion of their training, these are the young boys and girls who will lead creative research groups in R&D laboratories, defence, industry, atomic energy, space, educational institutions, etc. I strongly believe that it is not only important to identify talent but it is far more important to nourish and nurture the identified talent. The programme is aimed at producing peaks of excellence, leaders in science, attracting and stimulating bright young students all over the country. This is sought to be done by devising a five year integrated programme leading to M.Sc. degree available at either newly created institutes of science or at the best university departments in the country for producing highly motivated and trained scientists confident and capable of tackling problems in the rapidly advancing frontiers of their chosen disciplines. The product is expected to be a thorougly disciplined professional with detailed knowledge and understanding of his own subject and having an awareness of current and future trends in other disciplines and appreciative of the social milieu in which he would practise, its socio-political structure, its resource and growth problems and their solutions.

The input to this programme should be highly selective based on performance in a well designed national competitive examination. Admission to this programme should be based on no consideration other than merit and it should be open to the students who have passed the higher secondary examination from any place in the country. Initially, the admission to this programme should be limited to 700 students, so that when the programme becomes fully operational, there would be in all 3500 students constituting 0.5 percent of the students seeking admission to the first year of the three years science degree course. The programme may be gradually expanded to include roughly 1500 students per year.

This will be an intergrated science education programme leading to M.Sc. degree. An attempt will be made to integrate the programme in time as well in terms of disciplines. Emphasis will be laid on the unified structure of science as man's quest to understand nature and natural laws and the utilization of this understanding for the benefit of the people.

Having ensured selectivity and having identified talented students, it is essential to place them in a creative atmosphere in contact with highly creative, eminent and dedicated persons. They should see sparks of creativity in action and feel inspired and excited by scholarship and dedication. Two alternatives can be thought of for such institutions. One of the obvious possibilities is the setting up of 3-4 institutes of science on the pattern of Indian Institute of Technology. The other possibility could be to use the existing infrastructure of the best university departments in each of the science subjects. Much can be said in favour and against these possibilities, but the basic issue is that these selected students should be placed in a highly creative atmosphere. All the students will have a common course in the first year, so that all of them are exposed to basic physical concepts, mathematical techniques and to the excitement in life sciences. At the end of the first year, the students would opt for the discipline of their choice. Enough flexibility should be in-built into the system so that migration from one discipline to the other is permitted after satisfying certain qualifying criteria. The students will have experimental work which will include open ended exploratory experiments, project work and opportunity to participate in research programmes. It is perhaps possible to comment on the programme as being elitist and discard it but let me emphasize that it is not elitism by virtue of birth or economic, social or political consideration but is purely based on academic attainment. We must respect elitism in academics and indeed attempt to promote it, for Einstein, Bhabha, Raman, Rutherford are not many and cannot be compared with the ordinary.

We might consider the question as to how bright students would be attracted to this programme. I believe that those who pass the National Competition test which is aimed at selecting top 0.5 percent of the students at the higher secondary level and who are exposed to the creative atmosphere, flexibility of course and the excitement of science will not find it difficult to obtain a job comparable to one that a student after pursuing professional courses hopes to get and should therefore naturally be attracted to the programme. However, the problem is more psychological than real. The prospective bright students and more so their parents need to be assured of good career prospects. In this context, I would recommend that CSIR, DAE, ISRO, etc. should come forward to assure them that if they do well in the 5 year M.Sc. programme, they would be absorbed in these agencies at Scientist-B level. In the process, the agencies will be able to get the cream of young boys and girls suitably trained, and, in turn, it will remove the psychological barrier in the minds of children and their parents.

The second tier approach will involve selecting some 100 colleges, roughly 2-3 colleges per state. These colleges should be suitably funded to enable them to improve laboratory, library and other infrastructural facilities and to appoint suitably qualified and experienced staff. The colleges should be given autonomy to frame courses, conduct examinations and award degrees, under the overall guidance and supervision of a central steering committee consisting of creative research scientists in various disciplines. The admission to these colleges should also be based strictly on merit determined through a competitive examination held in the state concerned. These colleges would offer a large menu of courses and flexibility to the students to offer courses of their choice. These colleges will also provide first year course in such a way that all the students would be exposed to the basic physical principles, mathematical techniques and excitement in life sciences. The students will also be made to undergo workshop practice, glass blowing, electronic circuitry and computer programming. In the second year, the students would branch off to their respective streams.

These two programmes will account for nearly 16 percent of the students enrolled for undergraduate science courses.The remaining 84 percent students will pursue their educational programmes in affiliated colleges. The deficiencies in these colleges in terms of adequately qualified staff, ill-equipped laboratories, demonstration experiments are sought to be overcome through the use of electronic media, particularly the video system. It is suggested that eminent teachers, researchers, and creative minds will be invited to prepare video tapes of lectures addressed to course content. The video system is ideally suited to demonstration of experiments which the students in these affiliated colleges would never be able to see and enjoy the excitement of science. In a sense, through these video tapes, one is attempting to take the best teacher to the doorstep of a student even in the remotest village. In addition, the colleges will have a system of counsellors, so that a student after seeing a video tape can go to the counsellor to get his doubts cleared. I realize that it is not possible to replace a good teacher, but may I state that there are no good teachers to replace. The video lactures are intended to enrich the educational system in colleges and should be of the type and quality of Bragg's lecture or Feynmann lectures.

Besides enriching the formal education, these tapes will also provide the possibility of non-formal education. Far more important than this is the possibility of carrying out by implication teacher training. An inexperienced teacher on seeing these tapes would understand how a concept can be handled.

With an increase in enrolment in higher education, there has been a steep rise in the educated unemployment. We have reached a staggering figure of 20 million educated unemployed out of which science graduates account for a substantial fraction. Unfortunately, there is no proper manpower planning, which has resulted in a serious mismatch between the output of higher education and the demand for these people. We have reached a paradoxical situation wherein we have a large number of educated unemployed and yet we do not find enough well trained and skilled masons, welders, fitters, electronic technicians, medical instrument technicians, etc. Boys and girls having passed the higher secondary examination knock at the doors of the university to seek admission to degree courses irrespective of the fact whether they are qualified to puruse the course. It is time we consider the problem in all the seriousness it deserves and open out new channels and diversify our educational system. If social pressure is for a degree, the least that one can do is to provide for short term technical courses, such as that of electrician, radio and TV repairing, motor winding, computer maintenance, etc. Simultaneously, with the degree course, a student should take every year during three years of his stay in the college a short term course in these trades, so that when he goes out of a college with B.Sc. degree, he also has three diplomas in the trades for which there is a demand. Thus, even if he does not get a job on the basis of his degree, he could stand on his own. This will not only reduce pressure of the educated unemployed but we will have in turn better trained and qualified tradesmen.

In the end, I would briefly refer to the movement we are proposing to launch in Maharashtra and this concerns the setting up of what I have chosen to call exploratories. Most of us are born scientists, endowed with inquisitiveness and a strong desire to explore, to question and to discover. But many, perhaps most of us, lose our early enthusiasm for science when it becomes a formal subject, difficult and half understood, appearing to come as ex cathedra from great minds of the past to whom we find it difficult to relate. But these great minds usually experimented and thought about things which are apparently very simple and similar to and in some cases identical with the experiments that the visitors can not only see but also explore in the exploratory.

Today, we all live in a world of science and to be ignorant of science is to be ignorant of a large part of life. It is like almost going through life without one of our senses. Such ignorance puts us amongst the hard of understanding in most matters which concern twentieth century men and women as well as being a big handicap in finding interesting and rewarding employment. Many eminent men and academic bodies have expressed their concern about the public understanding of science and it is hoped that through such explorations, this understanding could be greatly improved and science projected as a result of rational thinking and exploration rather than magic.

Besides making general public more aware about science, the exploratory will serve another very important function and that is to arouse the curiosity among children to experiment and explore the basic concepts in various branches of science before they are spoiled by formal education. It is essentially through spirit of enquiry and an urge to know and to understand that science has grown. Indeed exploratories will attempt to provide the children to play and to explore which facilities in fact should be provided by schools and colleges, but are unable to do so because of large numbers and poor infrastructure. To an extent this deficiency is met in foreign countries by giving children mechanical, electrical and electronic toys. It is almost impossible for parents in India to provide their wards with such expensive toys. Exploratory intends to provide these facilities in much more organized and systematic manner with a view to promoting the spirit of enquiry in the minds of children and exposing them to the excitement of doing science. Hopefully, the joyful experience the children will get in the exploratories will prompt them to do science and inculcate in them scientific method and attitude.

Exploratories are fundamentally different from museums, and are essentially places where visitors are helped to explore basic scientific principles and enable them to see how such basic concepts are at the root of some major discoveries or technological advancement in the form of devices and artefacts. Visitor's interest is thus shifted from antique collection in the museums to interactive exploratory exhibits and possibly to futuristic presentation. Exploratories thus include exhibits with which visitors can play, explore and understand.

Lastly, we may touch upon process of science, that is, the scientific method which should become an inseparable part of education, whether it is of liberal arts, social sciences or natural sciences. There are many things that science has contributed; some of them have been of immense benefit to mankind, but quite a few of them have brought mankind to the precipice of disaster. A more important and everlasting contribution of science is the way science is done, the way one explores nature, the way one analyses situation and makes objective assessment. The history of science is a fascinating and glorious chapter in the book of human achievement. It is deeply embedded in man's quest to understand what happens around him. Science as we understand today has an idelible imprint of the Greek thought and its early struggle against religious dogma which gave science total objectivity. Dogmatism of any kind whatsoever is totally inadmissible in science. Dogmatism is subjective and its ultimate basis is personal prejudice or belief and not reason. Dogma is personal but science is communicable. The cooperative enterprise of science, because of its total objectivity, has been astonishingly successful perhaps more than any other enterprise of man. Scientific truth is open to all who wish to see. It is not restricted to the privileged few as their personal treasure. The objectivity of science has its origin in the scientific method based on keen observation, generalization of these observations, prediction of certain situations based on these generalizations, confirmation of these predictions by experiments, modification of generalizations on the basis of experiments and so on. This rule of the game is universally accepted and it is this tacit adherence to the rule that has made science totally objective. By scientific method, we do not mean simply the knowledge or the practice of science but the capacity to separate the factual and verifiable from the more emotional and traditional sentiments. In this characteristic method, we can distinguish two aspects: that of rationality and realism, that is, the ability to sustain by arguments and appeal to common experience. Science is autonomous and integrated. It is integrated within the domain of natural phenomena and if one exception to an established scientific truth is made the whole edifice of science will collapse. Science is autonomous in the sense it does not brook any interference within its realm whether human or divine. However, we recognize that the domain of science is expanding and today what does not lie within its purview of understanding may come into its fold tomorrow or the day after, but when it comes into its fold, it will become an integrated part of the same objective science. Thus, today we do not understand the nature of sub atomic particles, but we do hope to understand it some day or the other. At the moment, it is beyond science. However, it is knowable. Science thus recognizes that there is beyond science but cannot for its own self-preservation conceive of anything that is anti-science. This whole philosophy of science has been remarkably successful in unravelling the mysteries of nature and has indeed provided a new approach to deal with situations, to solve our problems. This scientific method and the scientific temper should be the basis of all education and every student, whether of liberal arts, social sciences or natural science, should be disciplined in this scientific method. Science makes available a large number of options; however, it is scientific method which enables one to make the correct choice.

In a sense, therefore, just as education in science is important for economic development, science in education is perhaps more important for social transformation and national reconstruction.

Programmes of Excellence and Equity in Science Education in the University System

S.K. Khanna

India, with its vast cultural and economic diversity, presents a wide spectrum of regional imbalances in the higher education sector. The statistics for the years 1971 and 1981 show that the proportion of people with post-secondary level of education in the age group 25 and above was only two per thousand in 1971 and four per thousand in 1981. The estimated proportion during 1988-89 is only 5.8. India has the third largest scientific and technical manpower, still the overall proportion of such manpower is six per ten thousand population compared to 70 in South Korea and 115 in USSR and 190 in Japan. This is a matter of concern when we review growth rates in S & T education.

A question is also raised about the expenditure on higher education sector and the priorities relating to elementary and adult education programs. It becomes more significant to consider the issue when a choice has to be made to provide opportunity to one person for professional education as against imparting elementary education to 100 children. The question of equity has to be examined in the perspective of all development sectors in comparison with the education sector and likewise within education sector, the competing claims of higher education and other levels of education. The specific concern of the share and equity for science education would make the issue more complex. All these elements are interwoven and they all interact and affect each other and we need to make a delicate balancing.

The equity can concern itself in terms of the region, gender, economic status and many other factors. We could on one hand consider the educational attainment of the population as a prime factor for the well-being of the country and on the other hand, we need to bring societal benefits and their equitable distribution among the people living in different parts.

Regional imbalances in higher education

There is a large variation in the disposal of educational opportunities at undergraduate level. The national average enrolment is 431 per one lakh of population and the lowest enrolment is of the order of 234-238 as in the States of Orissa and Bihar and at the same time Chandigarh and Delhi have enrolment of 4339 and 1396 respectively. Such variations exist between urban and rural population and enrolment varies by the scale of 40 to 3 per thousand population.

Factors which influence equity include: the system of secondary education which supports chosen few for higher education; medium of instruction at the school and university level; economic and social status of the families and unfortunate rate of drop-out among the weaker sections due to inadequate backup.

Thus, equity in opportunities for high quality education needs redressal at the school level itself, since this influences the system of higher education to a great extent. Educational planners and decision-makers therefore are caught with the dichotomy, as to how much correction be made and still maintain the necessary thrust of excellence. Equity is the ideal of social justice and it would remain abstract and virtually meaningless without a concrete and specific strategy. Question is how to chart the most effective and efficient path through the 'demands and supply' status of the facilities and opportunities knowing fully well that many of the options and opportunities are incompatible and also knowing that any shift in emphasis would have chain actions.

Promotion of equity

Perhaps, the most perplexing problem to the educational planners is to decide whether to follow or to lead the society. One must appreciate that science education has both social and corrective roles. It is through science education that we can come out of the conventions and beliefs which have caused damages including inequality and we can bring social change to reverse the conflict between the different sections of the society through science education. Science education can help us to promote human relationship. We know that the long term national development and stability, and satisfaction can best be achieved by quality and the spread of science education throughout.

While we discuss equity, it is necessary to take note of our dimensions of the higher education system. We have over 164 universities and over 6,500 colleges providing student enrolment of 3.5 million for undergraduate and 0.5 million for postgraduate studies. If equity has to be considered for social justice, one may have to further expand this huge structure, because even now the higher education opportunities are available only to 10 percent of our youth population. Equity would also demand minimum levels of facilities for teaching, research and examination and we would need substantial investment in doing so. In a small measure, University Grants Commission (UGC) has taken steps to reduce regional imbalances by way of providing funds to the educational institutions located in backward areas and through schemes for supporting education for women as well as for under-privileged section of the society.

About 40 to 45 percent of the development grant in a plan period which is 4 to 8 lakhs of rupees for individual college and 100 to 150 lakhs of rupees to the university/institutions, is used for building up of laboratories and other facilities for science education. Arising out of New Educational Policy (NEP) 1986, the UGC took measures for providing equity in science education by giving special grants to all universities under the scheme of removal of obsolescence.

Development of science education is closely linked with science educators and unless the science educators are motivated, no purpose is served, because the ultimate aim is to motivate students who could use the education experience in their day-to-day life. To motivate educators (teachers) and also to inculcate in them scientific temper, during Seventh Plan, ASCs

(Academic Staff Colleges) have been commissioned. These institutional arrangements provide necessary training to the teachers in teaching methodology and help in upgrading the subject knowledge.

Parameters to judge excellence in science education

When we talk about excellence in science education, many questions arise, particularly the criteria for judging excellence. What are the parameters responsible for promoting excellence? Should one judge excellence by counting number of National and International awardees? Should one rank team work in promoting excellence higher than individual brilliance?

There cannot be one single answer to these questions. We know that the level of excellence cannot be judged by one or two positive answers, but there are several inter-related aspects. The pursuit of excellence in science can best be achieved through total commitment to the cause and adopting appropriate strategies on the following aspects:

- Educational curriculum
- Identification, training and evaluation of manpower
- Environment and infrastructural development
- Linkages between institutions and development sectors, and
- Science information system and awareness.

Education curriculum

Curriculum is a single important element of the educational process and curriculum innovation is a must. It has been said that curriculum as is conceived now is made up of a few topics in narrow areas, which leaves much to the imagination of the students and the teachers. On the other hand, it has been argued that there is a shift in the emphasis from "Learning" to "Teaching" and therefore curriculum design needs to be carefully made to allow the student to learn in his own way instead of pursuing the exposition of a discipline. However, the case studies and the demonstration experiments in science education should always relate to solutions of real problems which exist around us. The UGC has taken steps in this exercise with emphasis on curriculum redesigning in such a way that the students of basic sciences can have access to knowledge of social sciences and *vice-versa*.

In the recent years, UGC, through the curriculum development centres established in the university system has generated model curricula through the interaction of large number of academics and others and these recommendations have been sent to all academic councils of the universities. It is hoped that this effort on the part of the UGC would influence the higher education system significantly. This process as mentioned earlier has to be a dynamic one.

One may appreciate that poor quality of undergraduate education will lead to poor performance of the university system both in regard to teaching and research programs with this in view, the working group on higher education for the 8th Five Year Plan, has stated that there is a need to identify a minimum of 100 colleges in the first instance to upgrade them as colleges of excellence where under controlled conditions, quality of education can be monitored as an experiment. There are arguments which also favour selecting a few talented students to enter into such colleges, so that all dimensions concerning the quality of education at undergraduate level could be carefully analyzed and understood. One must accept the fact that an excellent super-structure cannot be built unless the foundation is properly strengthened. The college sector is direct feeder to the university system and ignoring it would never bring dividends of excellence. It is our firm belief that any resources spent on university sector would go waste unless the college sector is strengthened.

Identification, training and evaluation of manpower.

Another question which needs to be examined relates to the manner in which the higher education institutions declare the students 'fail' or 'pass'. Probably the manner in which this process is conducted cannot be trusted as a proper instrument to judge the complex human intellectual and personality profile. It is realized by the academicians that 'objective' as well as 'subjective' assessment by the teachers based on prolonged evaluation of an individual can alone yield profile of the competence of such individuals and this could serve as an appropriate measure of excellence.

One cannot imagine continuation of excellence amongst students and the excellence of the institutions without the excellence among teachers. We need to come out of our traditional way and participate in the total exercise with full vigour and enthusiasm. In order to promote excellence among teachers, particularly with ever- expanding knowledge, the role of training and retraining cannot be over-emphasized. To this end, the UGC has established 43 Academic Staff Colleges in the university system. By 1989, nearly 16,000 teachers have participated in these programs. About 70 percent of the eligible colleges have been assisted during this Plan period by the UGC under Faculty Improvement Program through which the college teachers have been financially supported for acquiring higher research qualifications. Over 500 college teachers have received travel grants during the last two years to help them to participate in the international conferences abroad.

University and college teachers also undertake research projects sponsored by a number of national agencies, including UGC. On an average 400 to 500 minor research projects and 150 to 170 major research projects are awarded by the UGC in science subjects every year. It is heartening to note that the Department of Science & Technology (DST) during 1979-83 approved 242 projects under Science and Engineering Research Council (SERC) of which 110 were received by the teachers of the university sector and similarly out of 368 projects approved under General Research Scheme (GRS) during the same period, 178 were received by the teachers of the university sector. Such levels of award could be one indicator which may reflect the excellence of the teachers and performance of the institutions.

Excellence in research depends on identifying talents and encouraging them to pursue research and providing them necessary inputs. For this, the UGC in collaboration with Council of Scientific & Industrial Research (CSIR) selects every year Junior Research Fellows through an All-India Test. Research Fellows are encouraged to undertake research anywhere in the higher education system and research laboratories and other institutions of national importance.

Besides this, there are schemes like Career Award, National Associateships, Emeritus Fellows through which teachers with proven ability of research with sponsorship of the UGC avail exclusive time slots for

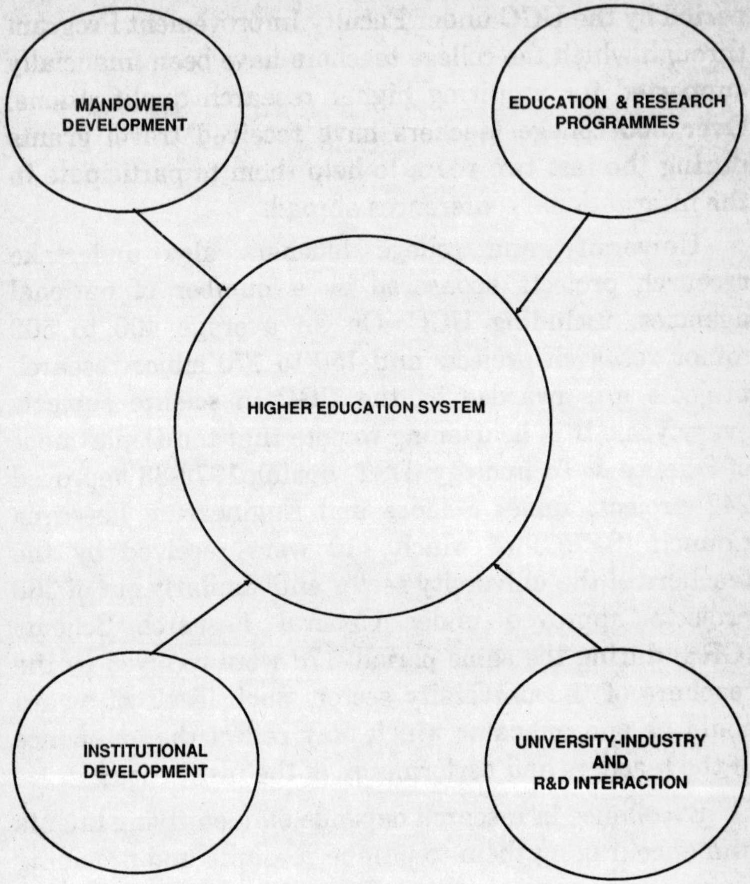

Major Inputs for Excellence in Science Education

FIG.-1

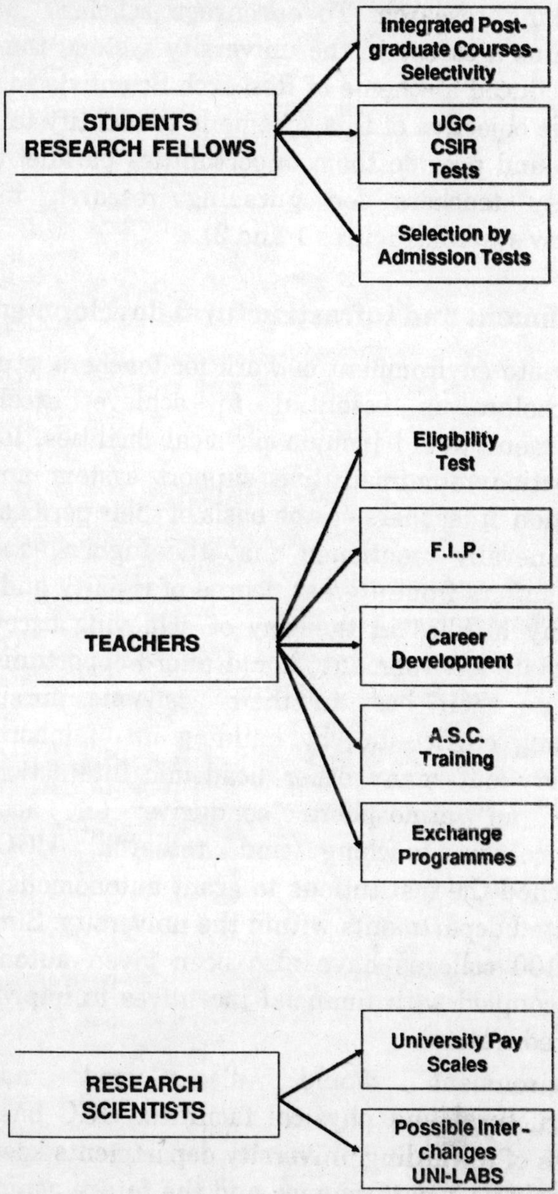

Excellence - Manpower Development

FIG. - 2

undertaking research. To encourage scholars to take research as a career in the university system, the UGC has introduced a scheme of Research Scientists in 1983. The basic objective of this scheme is to identify talented scholars and provide them opportunities parallel to the university teachers for pursuing research in the university system.(Figures 1 and 2)

Environment and infrastructural development

Appropriate environment of work for teachers, students and scholars is essential to achieve excellence. Environment would include physical facilities, level of participation, administrative support system and the recognition of teachers on the basis of their performance. It is generally mentioned that the higher education system suffers from a great degree of rigidity and there are many hurdles in the way of achieving excellence. Appropriate environment should afford opportunities to all to do their best in their academic pursuit by innovating curriculum, by building up of laboratories and carry out many other academic interactions. To provide an atmosphere conducive for achieving excellence in teaching and research, UGC has approached the institutions to grant autonomous status to selected departments within the university. Similarly, about 100 colleges have also been given autonomous status coupled with financial incentives to improve the college education.

Environment would also need adequate infrastructural and physical facilities. UGC has many schemes of rewarding university departments keeping in view their past performance and the future aspirations. At present, about 170 departments are participating in Special Assistance Program which was initiated in 1964.

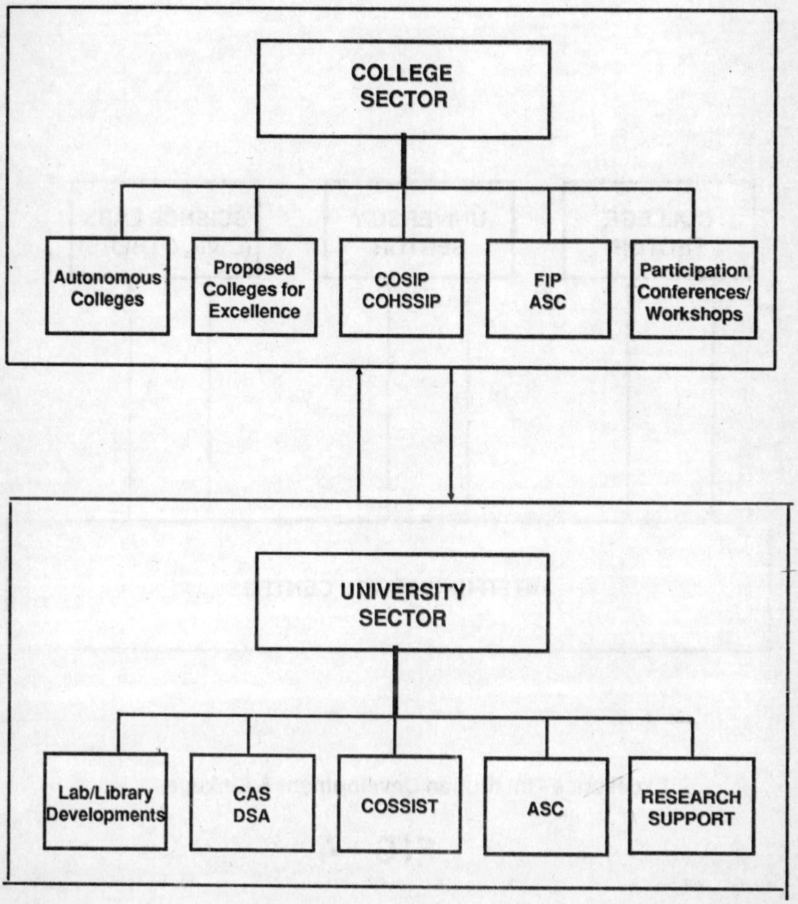

Excellence - College - University Linkages

FIG. - 3

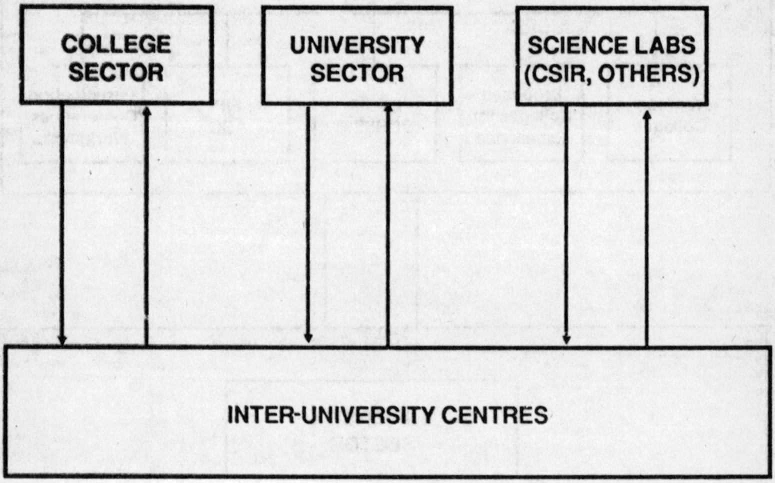

Excellence - Institution Development & Linkages

FIG.-4

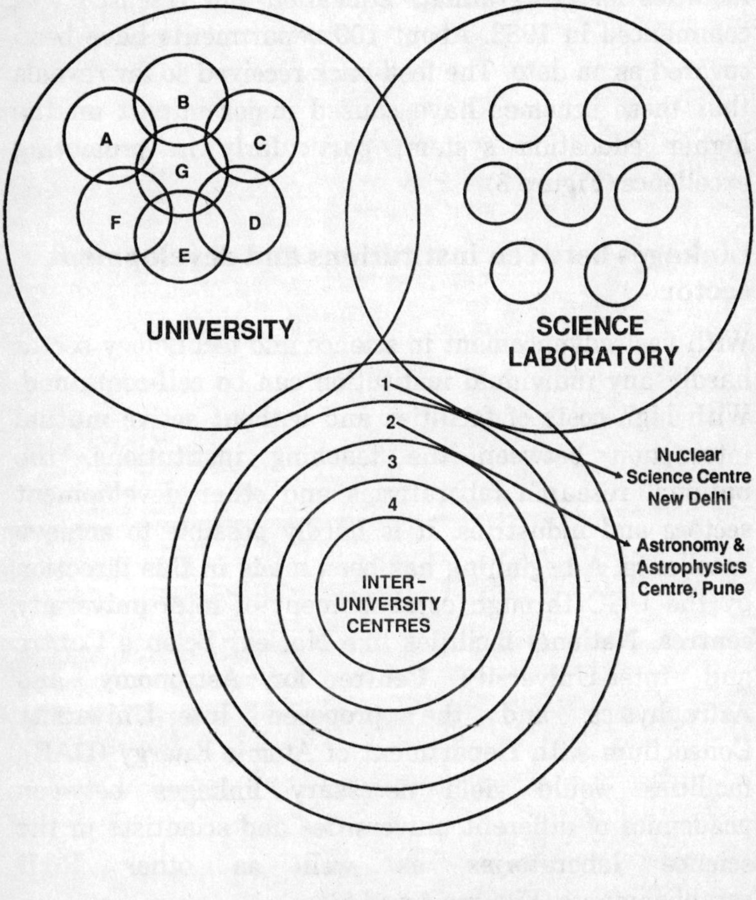

Interaction for excellence and growth

FIG. - 5

Selective support under the Committee for Strengthening of Infrastructure in Science & Technology (COSIST) schemes for strengthening the infrastructural facilities for postgraduate education and research was commenced in 1983. About 100 departments have been covered as on date. The feed-back received so far reveals that these schemes have caused major impact on the higher education system, particularly in promoting excellence.(Figure 3)

Linkages between institutions and development sector

With vast advancement in science and technology sector hardly any individual institution can be self-contained. With high costs of facilities and without active mutual interaction between the teaching institutions, the national research laboratories and other development sectors and industries, it is hardly possible to achieve excellence. A beginning has been made in this direction by the UGC through establishment of inter-university centres. National facilities like Nuclear Science Centre and Inter-University Centre for Astronomy and Astrophysics and the proposed Inter-University Consortium with Department of Atomic Energy (DAE), facilities would yield necessary linkages between academics of different universities and scientists in the science laboratories as well as other R&D establishments.(Figures 4 and 5)

Science information system and awareness

A very important feature of modern time is the expansion of knowledge. It influences development, and helps to produce new goods and services, shapes international cooperation and in fact, gradually, changes

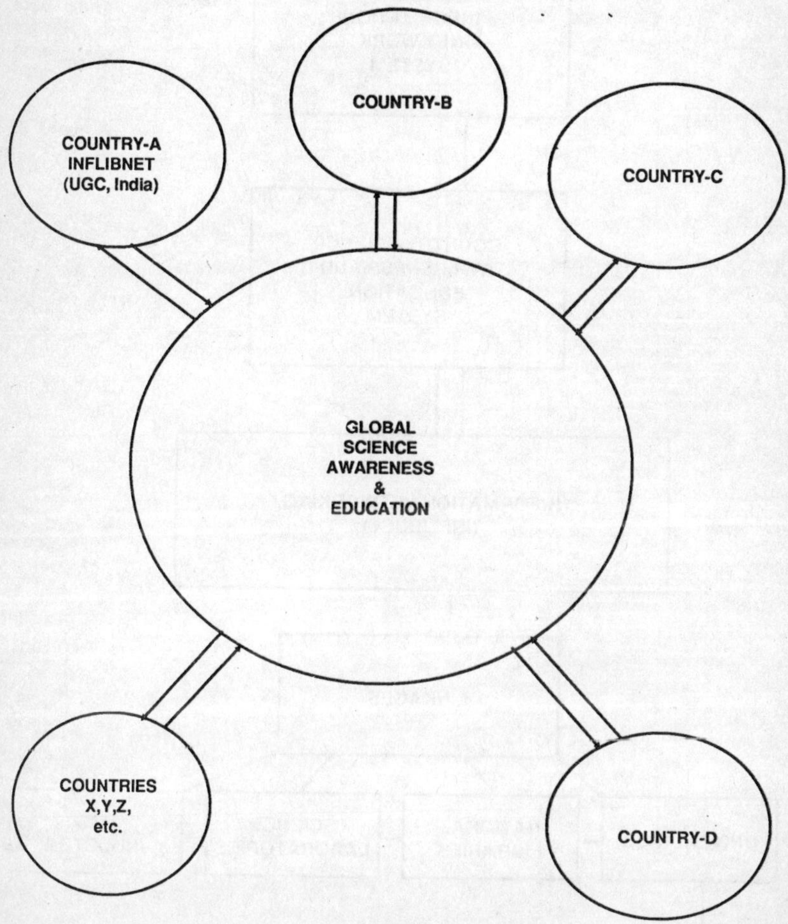

Excellence - Global Networking of Science Information

FIG.- 6

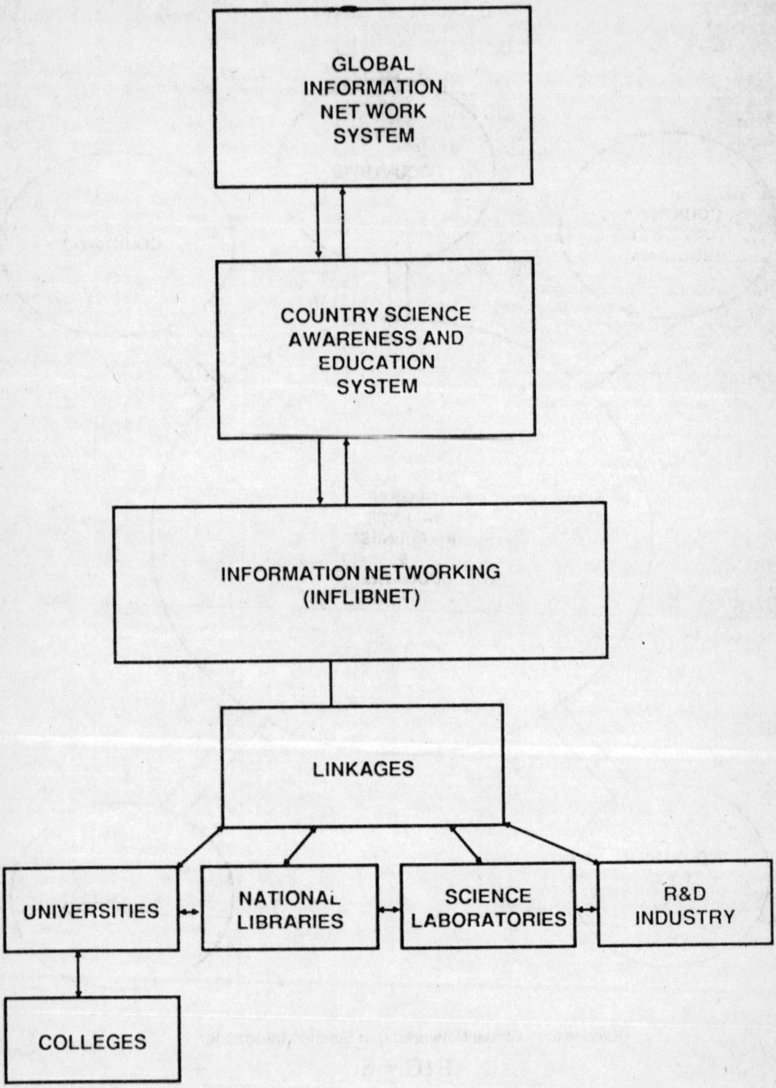

Schematic Frame of Global Science Awareness and Education

FIG.-7

the world order. Access to science information is shrinking in our country. This is due to resource constraint and steep escalation in the cost of publications. Most of the libraries have inadequate resources to subscribe even to minimum number of publications. Against this background, we took initiative of preparing a blue-print for establishment of a network through modern technology of information and communication system. The Information & Library Network (INFLIBNET) project seeks to link the library resources of all universities, institutions of higher education and research laboratories. It is hoped that a beginning in this direction will be made during the 8th plan period.(Figures 6 and 7)

Equity and excellence need not be considered in isolation. It is a question of balancing them keeping our national priorities in view. The major factor which prohibits promotion of equity and excellence together is paucity of resources. The total expenditure in Science & Technology is about 1.2 percent of the G.N.P. and we spend less than half of this amount in higher education, including Science & Technology education. We know that our higher education sector provides the entire scientific and technical manpower to all development sectors. It is hoped that linkages of higher education with the development sectors and industries would perhaps help in mobilizing additional resources. This would in turn help us and the educational planners to take measures for promoting equity and excellence together.*

---

* *Thanks are due to my colleague, Dr. P.B.Tripathi, for his consultations and furnishing me data for this paper.*

Education within the Framework of People's Science Movement

Vinod Raina

An understanding of the development process in post-independent India would require digging into the colonial roots. Thus, Marx's famous observation about the dual nature of colonialism has found a curious echo in the body-politic of independent India. He observed: "*England has to fulfil a double mission in India: one destructive, and the other regenerating — the annihilation of old Asiatic society, and the laying of the foundations of Western society in Asia .... The political unity of India ....imposed by the British sword, will now be strengthened and perpetuated by the electric telegraph. The free press introduced for the first time into Asiatic society .... is a new and powerful agent for reconstruction .... Steam has brought India into regular and rapid communication with Europe .... and has revindicated it from the isolated position which was the prime law of its stagnation .... The millocracy (British industrialists) have discovered that the transformation of India into a reproductive country has become of vital importance to them and, to the end, it is necessary above all to endow her with means of irrigation and internal communication .... All that the English bourgeoisie may be forced to do will neither emancipate nor materially mend the social condition of the mass of people, depending not only on the development of productive powers, but on their appropriation by the people*". In fact, it is widely accepted that the economic and political domination of the colonial power was enabled precisely because of its superiority in both technological capacity and in the realm of 'progressive' ideas — an euphemism for rationality and modern science. As such, colonialism played a 'positive' role in pinpointing the inherent weaknesses of traditional Indian society[1].

Such a reading of colonialism crucially influenced not merely the processes of State intervention in independent India, but the response of dissenting intellectuals as well. Heavy industrialization was the means by which Nehruvian economics sought to transform a 'backward' society. Subsumed in this economics has been the naive hope that increased productivity would automatically result in an equitable distribution of the produce and profits. The crucial pillar on which this whole thesis rested was a massive input of modern science and technology (S&T). Essentially, this hope and thesis continues to dominate the Indian body-politic. In fact, it has been pursued much more vigorously in recent years, even though the resultant capitalist growth mediated by neo-colonialist multinational industry and multilateral funding agencies has evoked widespread discontent throughout the country. Ironically, Nehru's temples, the large scale irrigation and power projects are being vigorously opposed, be it Tehri, Narmada or Koel Karo, because of questions of unequal sharing of benefits, human costs and serious effects on the already depleted natural resources.

Marx's observations were not confined only to issues of economics and technology under colonialism. They went further in exploring the very nature of ways of thinking within the village communities and he asserted that: ""*we must not forget that ... these idyllic village communities... had always been the solid foundation of Oriental despotism, that they restrained the human mind within smallest possible compass, making it the unresisting tool of superstition, enslaving it beneath traditional rules, depriving it of all grandeur and historical energies..... We must not forget that this undignified, stagnatory and vegetative life ... evoked..... aimless, unbounded forces of destruction..... that these little communities were contaminated by distinctions of caste and slavery, that they subjected man to external circumstances instead of elevating man.....*" It is of immense interest to note that these observations have come under a great deal of strain in recent times through debates on sustainable development which seem to have resurrected the relevance of village communities. This has given impetus for attempts to work out the political economy of Gandhi's beliefs and assertions about the relevance of rural communities.

However, more relevant to the theme here, the emancipation of the mind found an expression in the Nehruvian thesis of 'scientific temper', a term least understood but much publicized. In effect, it has remained a mere rhetoric empty of content. The proof of the emptiness of this rhetoric is evident from the strain on the science-society interface resulting from the Bhopal gas disaster. Here is a glaring example that has gone abegging in defining the links between toxicology and medication, environmental hazards and citizens' rights, and most important, scientific evidence and legal issues. In the absence of even a rudimentary 'scientific temper' within institutionalised science in the country, the mass murder of over 3,000 persons and long term health effects on over two lakh persons have been priced at a few hundred crores, with no liability to the offenders. In a sense, none of this ought to be surprising. A developmental pattern motivated by profit and capital, with the concomitant of a 'neutral' and 'value free' science, is essentially inimical to a critical mode of thinking that may question the values and policies of such a framework. Just as a profit motivated production process demands a disciplined, controlled and, hence, a servile workforce, so does it demand and ensure an uncritical mass of population that serves as a market, willing or unwilling, for the goodies. After all, where will the profits come from?

The production process, however, does require a fairly large human force, but they need only have skills of a technical and managerial kind; an overcritical attitude would in fact be a threat to the whole process, and hence to the body-politic. Little wonder then that the basic educational infrastructure has remained relatively unchanged in approaches to content, method and pedagogy from colonial times. One may, therefore, grossly conclude that whereas the educational infrastructure during the colonial times was shaped essentially to service processes relating to the appropriation of the colony's natural wealth, in free India, the emphasis has broadly shifted to servicing the appropriation of surpluses. Such conclusion is indicative of the fact that the education infrastructure in a colonial and a capitalist system have a common residue; only the beneficiaries are, respectively, the colonisers, or a small segment of fellow countrymen and foreigners mediating via multinationals.

This then is the basis of the disjunction that has gradually emerged in the country, between the masses in general-people-and the minuscule rest. Media and educational onslaughts strive to make people accept that all this is 'progress' and 'development' and at the heart of such progress and development is the particular use of science and technology.

The role of a *People's Science Movement* (PSM) is therefore, naturally defined within this framework as not merely that of communicating and simplifying science (though that is essential) but also questioning each aspect of such progress and development, in particular the S&T issues involved, and intervening wherever necessary with the active participation of the so-called ordinary men and women. That is what spreading 'scientific temper' ought to mean.

Assumptions for critical thinking

Having indicated a role for the PSM, the immediate question that arises concerns the methodology of action, against the backdrop of economic and political conditions outlined. A related question often asked in this regard in many different ways may be summarised, in the context of education, in the following manner: 'Given that the educational process is actually determined by the overriding political process, what then is the point of advocating any alternative within the educational framework? The alternative must be sought for the body-politic itself'. The stock phrase is that education is at best a 'post revolutionary' activity. The question has also arisen specifically in the context of PSM — Science for social revolution or social revolution for science?[2].

The relevance of the question can hardly be denied and it must be admitted that a complete answer to the question is perhaps impossible on purely theoretical grounds and it is only through the combination of theory and practice that answers can be explored. However, it does seem necessary to indicate an approach in furthering the debate on this question. Such an approach is not being suggested merely for academic purposes. Instead, the approach is itself indicative of the lines of action PSM could follow. Also, this question continues to create a nagging unease to practising groups and hence needs to be opened up for wider debate. Instead of inventing a fresh argument, in what follows, we shall briefly outline some aspects of Gramsci's ideas that seem relevant in the context[3].

The particular idea that seems to have the power to crystallise and unify various experiential observations is that of 'hegemony', a concept central to Gramsci's analysis of stratified society. The notion of hegemony is most familiar in political history and international affairs where it refers to situations in which one nation exercises political, cultural or economic influence over others. But Gramsci extended its reference to apply to relationships between groups, especially social classes. Hence, one social class can be thought of as exercising hegemony over other 'subaltern' classes. For example, in a capitalist society, the upper classes can be said to be hegemonic in relation to the workers, peasants, tribals, etc. In this formulation, hegemonic direction is by moral and intellectual persuasion rather than by the police, the military, or the coercive power of the law : rule by intellectual and moral hegemony is the form of power which gives stability and it is based on wide-ranging consent and acquiescence. For this to be so, every relationship of hegemony is necessarily a pedagogical relationship. Control of the subaltern classes is much more subtly exercised than is often supposed: it operates persuasively rather than coercively through cultural institutions, religious organisations, labour unions and other worker associations, schools and the press. Or, as Raymond Williams puts it, "so totally do the complex forces of hegemony saturate the whole process of living that the pressures and limits of what can ultimately be seen as a specific economic, political and cultural system seem to most of us the pressures and limits of simple experience and common sense. That is, we are persuaded that maintenance of the *status quo* could not but be in our own best interests".

What then would constitute a counter-hegemonic action? By counter-hegemonic is implied seeking alternatives to the *status quo* in areas such as science and technology, pedagogy, and culture. If one goes along with *Gramsci's ideas*, the assumption is that the ordinary people must learn the mental and related skills of the dominant ruling class. Hence it becomes clear that the counter-hegemonic task is one of education, in the broadest sense of the word. It is imperative, therefore, to include some general principles to the specific nature of such education. This is particularly so because of the lack of clarity normally faced by PSM groups in grasping the tension between the dominant culture and education and the so-called people's culture and eduction. To quote Gramsci, "it must first be shown that all men are 'philosophers' or 'scientists' by defining the limits and characteristics of the 'spontaneous philosophy' or 'spontaneous science' which is proper to everybody. This philosophy is contained in :

1. language itself which is a totality of determined notions and concepts and not just words grammatically devoid of content;
2. 'common sense' and 'good sense';
3. popular religion and, therefore, also in the entire system of beliefs, superstitions, opinion, ways of seeing things and acting which are collectively bundled together in the name of folklore".

The formulation, akin to Piaget's psychology, amounts to saying that untutored intelligence frequently manifests itself in conceptions, which are somewhat primitive (pre-operational) with reference to refined scientific and social modes of thought. 'Common sense' is a blend of 'good sense' and folklorist superstition. A positive people's action would, therefore, be concerned with the development of 'good sense', a state of mind in which common sense is exalted in its superstition and folklore and given a coherent unity through exposure to philosopher's philosophy, and scientist's science, with the assumption that every man or woman, is a man or woman of science.

Hence, we have the inevitable conclusion that the practice of counter-hegemony is the practice of such pedagogy that derives knowledge, science, history, etc., in the classroom or outside, from popular folklore and commonsense. In this way, the development of the component of good sense would not be a dominant, substitutive and exploitative enterprise, ridiculing, demeaning and stigmatising the common sense of the people, but an additive component which while preserving the common sense, sharpens it to make it equal to, or more than that of the ruling dominant class.

Implications for science education

From the general, we now come to specifics, in order to formulate concrete action points within the PSM framework. In particular, an attempt shall be made to analyze the implications for science education.

The school structure is an obvious focal point for initiating processes of critical thinking. However, the hope that the schools of independent India, unfettered by colonial considerations, would exhibit a radical departure in approaches to content, method and pedagogy is yet to materialise. Why is it so? In particular, why has the science curriculum remained so stagnant? The following account is taken, almost verbatim, from Anita Rampal's illuminating account of the evolution of the science curriculum[4].

It has often been felt, and assumed, that objective observation and analytic abstraction inherent in the method of science provides a liberating potential to develop deep social insights. It is remarkable that John Dewey should have such commonality of views with Gramsci when asserts that "Whatever natural science may be for the specialist, for educational purposes it is knowledge of the conditions of human action" and "... the functions which science has to perform in the curriculum is emancipation from local and temporary incidents of experience and the opening of intellectual vistas unobscured by the accidents of personal habit and predilection".

Despite these assertions, the curriculum has not really responded to such vision. What emerged then and has continued to provide the basic framework for school education is a model of 'pure' abstract science, in consonance with the demands of logical coherence as viewed by professional practitioners of the discipline, making Dewey lament that "....aloof from the practical needs of the mass of men, advocates of scientific education put themselves at a strategic disadvantage".

An interesting account of changes that took place in the British educational system at a time when the capitalist path had truly been laid in that country (nineteenth century) is of extreme relevance here on two counts. Firstly, it is situated in a time-frame when the political transformation in Britain was somewhat similar to what is happening in India now — a transformation from a feudal to capitalist economy. Secondly, it lays bare the origins of the cumulative thinking that has shaped the notion of school curriculum in science all over, including India. Two major alternative conceptions were available then : the 'science of common things' and 'pure laboratory science'. The former was initiated by an influential clergyman, Reverend Dawes, with the aim of instilling self-confidence and integrity of thought amongst the children of the poor, in an attempt to ultimately improve their moral and religious condition. The curriculum was consciously designed within the cultural context of the labouring classes and "....was no crumb of upper class education charitably dispensed". Prominence was given to the applied science such as mechanics and agricultural chemistry to provide familiar experience for the exercise of reason, speculation and imagination. The success of the programme and its imminent expansion to all elementary schools was viewed as a major threat to social hierarchy; it was feared that the classics-based education for the higher orders, which had insistently excluded instruction in science, would prove ineffective for the domination of 'scientifically' trained mass of the poor. Voices from various influential quarters rose to express the perceived social danger in imparting superior education to those inferior in position. Eminent scientists like Lyell and Faraday joined in explicitly stating that "from a political point of view, it is not only an unhealthy but also a dangerous state of things in some respects, that the material world should be very much better known by the middle classes of society than by the upper classes". A double-edged campaign backed by the newspaper *The Times* advocated the introduction of 'pure' science, fit for the 'superior' mind in the liberal curriculum for the higher orders, and, simultaneously, a halt to the practical scientific education of the lower orders, since it had become 'too good' and expensive for them and was "...too far above basic literacy (required) for Bible reading". As a result, science was soon removed from the elementary school curriculum on the grounds of inadequate funds and trained teachers. It was further stressed that a "....knowledge of common things was not to be obtained by the direct study of science, but through country walks, star gazing and domestic experience". Two decades later when science reappeared in the school it was in a form very different from that envisioned by the radical 'science for liberation' advocates. The goal of 'pure' science in schools, as expressed by its new supporters, was to inculcate habits of 'value-free' and 'disinterested' inquiry 'for its own sake' in order to prepare future scientists for universities. In the words of an eminent professor, "...teaching of science in schools was to be means of sifting out of the great mass of the people those golden grains of genius which are now too often lost amongst the sands of mediocrity". Science in the school curriculum thus became defined as a highly selective sieve to sort out those few who conformed to the image of the 'pure scientist'. Indeed, the mass of the populace, for whom it was a resource accessible only in theory, were effectively deprived of any meaningful education in science, owing to its intended abstraction and aloofness from matters of their everyday lives and belief systems.

One can clearly see the influence of such formulations for school curriculum right through to post-Independent India, until the present day. The irrelevancy of the primary and middle school curriculum content, particularly in relation to the rural and *adivasi* populations which form a large majority of total school going children, has been pointed out repeatedly over the years. But the overriding factor influencing each curriculum revision and change has been to accommodate more and more of specialist science in lower classes, with the justification that there is no other way to cope with the 'information explosion' than to push in more and more information into curriculum. Under this onslaught, all criteria that have a socio-political bearing or even such considerations as age-concept correlations, have been swept away. So, if it is necessary that the physics of atoms and molecules has to be pushed down to class 5, because more information needs to be accommodated in class 6 onwards, then so be it. Whether class 5, or, for that matter, even class 8 students can actually make any sense of the concept of an atom is no longer a consideration.

Together with prevailing pedagogical methods, the 'pure and abstract' information based curriculum constitutes an extremely strong repressive tool of people's minds. An authoritative and one-sided mode of communication mediated by the all-imposing teacher from whose lips flow pearls of information that is alienated from the entire life values, aspirations and concerns of the sullen recipient constitutes, nearly wholly, the school transaction. A counter-hegemonic action, therefore, has a dual role — in intervening in the curriculum and in the pedagogical, teaching-learning methods.

The Hoshangabad science teaching programme

With this background of the historical evolution of the science curriculum, shaped by dominant concerns and interest in a stratified society, we shall now examine an ongoing programme that has attempted to re-establish the Dewian vision of science education in a, hopefully, counter-hegemonic manner. Naturally, it has meant conflicts in various areas and at various stages. In what follows, we shall confine ourselves to discussing some of these conflicts rather than a chronological narration of the whole programme[5]. These provide further insights into directions that educational action could take within a PSM framework.

At a pedagogical level science education offers a straightforward methodology with far reaching effects — that of activity based teaching. Implementing such a methodology has been one of the main concerns of *Hoshangabad Science Teaching Programme* (HSTP). This seemingly simple sounding change, when practised, particularly in the more feudally oriented rural schools, has been shown to have the power to break down the usual notion of teacher-child interaction. Such breakdown includes the re-orientation of the classroom architecture where rows of 'discipline' and silent children sitting in neat rows are liberated into groups of excited and chattering bundles of energy and joy, engrossed in their activity at a level of 'indiscipline' or freedom, that is so essential for the transaction of critical knowledge.

Science at elementary school levels also offers the opportunity to generate information through experiment and activity, making it less binding to supply it through comprehensive texts. Thus, it is possible to bring a sense of discovery into action, and this has been the other major attempt of HSTP. However, the balance between the amount of information that can be generated and the amount that needs to be supplied, particularly in higher school is a matter that requires further delineation through practice. Even so, it may be contended on the basis of about seventeen years of cumulative experience of HSTP, that in areas where information generation is difficult and facts have to be supplied, a method that retains a critical mode is very much possible.

At the level of content, HSTP has worked under a constraint. Since the programme was initiated by choice in rural government schools, and expanded within the same system, the number has risen from 16 in 1972 to 411 in 1988. An operative condition has been that the syllabus should remain essentially the same as prescribed by the State Board, to ensure 'parity' with non-HSTP children. It would, of course, have been easier to overcome this constraint by opening one's own schools, but the major objective of the programme has been to attempt changes within the organised education system, rather than prove the viability of a few islands of excellence. However, a major area of confrontation, over the years, has been precisely around this aspect. For example, contrary to the prescribed syllabus, we have omitted chapters on atoms and molecules and chemical formulae, which we contend are too abstract at the middle school level. Similarly, we feel that the correct place to bring in the concept of mass is when one takes up Newton's laws of motion (in high school). The space provided by these omissions has been utilised for a much more thorough going treatment on learning skills and concepts of measurement, decimals, flowers, crops and vegetables in the immediate environment, genetic diversity in nature using the village environment as a laboratory, and elementary notions regarding chance and probability. And all this through the mode of activity and experiment. In spite of this, it has generally been felt that the curriculum could, and should, be made more and more relevant to the local environment. In this environment is included the social and cultural environment. For example, the attempt to use a simple and straightforward language is another point of confrontation, with the mechanical educationists and scientists showing grave concern over the simplification or omission of technical terms.

The sustained attack on HSTP regarding the omission of topics like atoms and molecules reveals the same tendency favouring 'pure' science as described in the previous section. Similarly, the insistence on a precise, 'objective' and dry language replete with technical terms is to be viewed within the context of the particular function of science education that considers school education primarily as a training ground to produce stock scientists and technicians. That science education could have a social function in creating avenues of critical thinking is implicitly denied by such a contention.

A major area of confrontation, and innovation, has been that of kit materials. An activity based method obviously requires a certain amount of such materials. But within the Indian rural school system, this is considered a luxury by official standards. Within the priorities determined by the overall economic and political considerations, a demand for such expenditure produces an incredulous response. The mention of a science kit possibly conjures up images of expensive laboratories, glittering glassware, winking digital lights and all the other science stereotypes many science institutions and media have established. Isn't science apparatus after all an expensive business?

It is a matter of despair that even after seventeen year of HSTP, such stereotypes should continue to dominate the thinking of large sections of scientists and educationists. Such thinking persistently reveals a shocking ignorance of science and its method, in particular as needs to be practised in elementary schools. There are a variety of ways in which such ignorance gets expressed. Consider the following statement: "in science precision and exactness is everything, so, apparatus and instruments must be accurate, and hence, expensive". Similarly, it is assumed, without question, that packaged instruments, more or less in the form of black boxes, is science apparatus, tinkering with it science, but opening it to see its insides is spoiling it and hence, taboo. Conversely, constructing apparatus from ordinary bits and pieces is at best a hobby and play, not *real* science.

The chapters on measurement in HSTP workbooks, called *Balvaigyanik*, illustrate how some of these notions have been questioned. Take 'exactness' and 'precision', for example. At training workshops teachers are asked to measure the same length, say that of a classroom, individually, and record their observations on a piece of paper. All these observations are then written down together on a blackboard. The teachers are horrified to see the variation and discussion ensues on the causes. Satisfied, they remeasure the same length, to discover that variation, though less, still persists. Naturally, they feel very uneasy, because their notion that science is 'exact' is shaken, because they find it impossible to define the length of the room exactly. This then is the take-off point for a vigorous discussion on the notion of exactness in science, the unavoidable use of statistics in defining the length of the room from a set of actual measurements and so on. This is what then gets transmitted to children, and acts as a motivation for the use of the scale for measurement, thereby continuously reinforcing this important skill. So, by actual practice students and teachers realise that no matter how careful you are, repeated measurements, produce variations which are ultimately linked to the resolution of the measuring instrument. So, 'how exact?' becomes a contextual question rather than an absolute notion. A carpenter needs accuracies of the orders of millimetres, whereas in measuring road distances, variations by a metre may be normal. To insist on millimetre accuracy for such measurements for normal purposes is not really scientific, but silly.

Now all this gets done using essentially the child's standard 15 cm plastic scale and a wooden half-metre rod. To build up such an understanding of measurement, it is unnecessary to invoke vernier callipers, screw gauges and the like, which are usually considered an essential part of school science, and cost much more. It is not as if children should not be exposed to these measuring devices at some time or the other. The point is that if they are not available for any reason, measurement can still be learnt with deep insights by alternative means. In fact a mere use of these instruments without bringing out the deeper questions is a fairly superficial exercise.

Other innovations in the kit involve the use of a single item for many purposes. A stirring example is that of the plastic cube. This one centimetre cube now costs about 10 paise and is extensively used in learning concepts like area and volume. But being of nearly the same density as water, it is also an adequate measure of 1 g of weight and is so used in many weighing experiments. Finally, in the interesting chapter on chance and probability, the same cube finds use as a dice.

A famous example, told and retold many times is about the *'babool ka kanta'* (babool thorn). Initially a dissecting steel needle, costing nearly a rupee, was part of the kit for biology activities. A school teacher found, in the babool thorn, a local and freely available substitute. The needle has since then been removed from the kit and the all purpose *'babool ka kanta'* dominates all activities requiring puncturing, impaling or dissection.

Such an atmosphere where innovations, substitutes and new ideas are constantly sought has ensured that many of these ideas come from children and teachers. In fact, the degree of innovation exhibited by the rural school teachers has been a great source of inspiration of resource scientists of the programme who come from such institutions as T.I.F.R, Delhi University and I.I.T.s. The used ball-pen refill is another example. Available easily in all schools from children and teachers, this throwaway object combined with a flattened paper clip has emerged as the best method making an axle for a variety of toys, games and apparatus. Similarly, ordinary buttons, sealed back to back under a candle flame, provide multipurpose wheels and pulleys, and are the basic resource tools for a chapter on machines.

The ingenuity of the teachers and children surprised many chemistry experts when it was found that they had discovered that purgative tablets (like *Vaculax*) contained phenolphthalein and could be used as an indicator for titrations. This discovery has spurred further activity in this field with teachers, children and resource persons combining their ideas in exploring flower petals-of hibiscus, marigold etc., as indicators, because it is observed that many of them change their colour on addition of acids or alkalies.

Apart from such innovations, the simple pedagogy employed in HSTP is that of field visits. Soil profiles, crops, pests, insects, flowers, life cycles of mosquitos and frogs and numerous such activities require no more than a friendly guide, who could be the teacher, the local *patwari*, *gram sewak* or health worker and an acceptance of the fact that more enriching and environment-linked learning situations exist outside the four walls (if any) of the classroom. The joy and enthusiasm and the liberated feeling on the children's faces can be infectious.

These innovations are however another cause for conflict and confrontation. A criticism expressed by 'experts' is that it is at best '*picchda vigyan*' (backward science). By confining children to the bits and pieces of the 'poorly' endowed rural environment, children are denied exposure to 'modern' science as reflected by spacecraft, nuclear reactors, winking and blinking instruments, computer, etc. Such criticism often finds expression in sentences such as "...Man has reached the moon and you are still asking children to use babool thorns". It is a matter of deep regret that 'good' science and all of education has got linked to the arrival of a computer or a video in a few schools of the country. That science or any form of learning is discovery, curiosity, activity, imagination, participation, manual skills and critical thinking is being swamped and over-run by incomprehensible and alienating technological gadgetry, thereby undermining the purposeful role such gadgets could have if properly incorporated into a sensible education programme. And since it is virtually impossible that such gadgets can reach each and every school of the country, their presence in a few (like *Navodaya Vidyalayas)* shall further strengthen the stratification process.

What about the costs? For a school with an average of 40 children in classes 6,7 and 8, that is for about 120 students, the initial cost of the HSTP kit is less than Rs 1,500 i.e. about Rs. 12 per child. The replenishment cost of consumables and breakage works out to rupees one and a quarter per child per year. These figures are borne out by sixteen years of experience and are not mere projections. Surely, such an investment is not outside the possibility of State Government budgets, in any case, they are a right of every Indian child.

An area of intense effort has been in teacher orientation and involvement. In fact, it is this area that comes closest to the Gramscian ideals of increasing the component of 'good sense'. Sustained effort through large-scale orientation and training camps and monthly meetings have by now involved and equipped about 1,800 teachers in all aspects of the programme, like book writing, kit improvisation, open book examination and conducting an activity based methodology. This has succeeded to the extent that many among them have become inventors of pedagogy rather than mere mediators of a pedagogy.

The level of confidence attained by many of the teachers is quite unheard of amongst the administratively exploited rural middle school teachers of Madhya Pradesh. The following is one such illustrative example. The resentment of the local political forces has found expression in diverse ways over the years, mostly unrelated to the teaching-learning process. However, one such onslaught took the form of the local ruling party Congress MLA (later a Minister) demanding in the state assembly the closure of the programme since it was teaching gambling (*satta*) to children. His outburst was against a question framed for the board examination by a group of local teachers concerning the unit on chance and probability. The question asked children to compute the probability of winning a bet on a two digit number commonly used in *satta* playing. The question further enquired of the children to deduce whether the whole enterprise was a losing or a winning proposition for the player. The MLA, totally unaware of the content or intent of the question, surmised that since HSTP stressed activity, the two together must imply that children were being exhorted to practice *satta* in and out of school. The question was messed up in the state assembly and ultimately it transpired that concerned teachers were asked to submit a written explanation. In a spirited defence, they maintained that *satta* was a social menace and the study of chance and probability provides a clear rationale to prove that it was eventually a losing proposition. This, they maintained, was a much better condemnation of the practice than all moral appeals. In particular, they questioned the contradiction in the official position, which on the one hand exhorts teachers to provide link between what goes on in the school and the society outside, and precisely when this is done, charge-sheets them. So forceful and logical was their reply that any notions the State may have had for punishing them quickly evaporated.

Though this was a strong expression of the teachers' acceptance of the values of science and society, and somewhat high profile because of the accompanying sense of drama, it wouldn't be an exaggeration to suggest that such an expression was not an isolated event and that in many subtle and low profile ways such re-orientation of attitudes and values has been widespread. That it will and should lead to further conflicts is clear, but in a sense, that is also a measure of achievement of the programme.

However, it is necessary to sound a word of caution regarding the 'romance' of conflict situations. Increasingly, many practitioners of educational and PSM action tire, over the years, of low profile non-agitational work and feel the need to switch over to overtly more radical paths. It may partly be because of the Gramscian assertion that the degree of stability of the *status quo* is reflected by the ability of the existing hegemony to accommodate alternative and counter-hegemonic cultural forces, neutralising, changing or actually incorporating them. Such a feeling could be a cause for practitioners to seek direct conflict situations because of the sense of adventure and self-importance they provide. This is not to deny ultimately the need or validity of situations, rather than the incident in isolation. In Gramsci's words, "resort to violence by a subaltern class is not a sufficient condition for establishing its own hegemony, this requires a profound change in mass consciousness" and "there can, and must be, hegemonic activity even before the rise to power, and that one should not count only on the material force which power gives in order to exercise effective leadership". Hence, the rubric, "Instruct yourselves because we shall need all our intelligence. Agitate because we shall need all our enthusiasm. Organise yourselves because we shall need all our power".

PSM groups require to reflect deeply on these issues, to understand and carve out a role, in harmony with other forces, like scattered intellectuals, not necessarily scientists, trapped in universities, medical colleges and other such institutions, ordinary people and in particular progressive political parties.

Notes & References

1. Ramachandra Guha: 'Modern Science — That Holy Cow', *Frontier*, April 27, 1985.
2. For example, at the *1st People's Science Congress, Cannanore,* February 1988.
3. Gramsci's own writings are to be found in the well-known *'Selection from Prison Notebooks'*, Lawrence & Wishart 1971. A critical reading on which this account is based is to be found in Harold Entwhistle's *'Conservative Schooling for Radical Politics'*, RKP, 1979. It should be made clear that the reference to Gramsci's views is not in order to invoke a sense of authority, nor is it meant to open up controversies within the Marxist framework regarding 'Humanist Marxism','Western Marxism', 'Euro-communism,' 'Base- Superstructure', etc. This is neither the place nor the occasion to get involved in such debates. It is just that these views seem to have contextual validity.

4. The account is to be found in Anita Rampal's paper *'School Science in Search of Democratic Order'*, Nehru Memorial Museum and Library, 1988. This paper, amongst other things, traces in detail the evolution of the science curriculum as shaped by social and political influences.
5. Many accounts of different aspects of HSTP are available, for example:*Science Today*, December 1977; *India Today*, July 1984; *Economic & Political Weekly*, No. 42, October 15, 1988, and *People's Action*, CAPART (forthcoming).

Information Support for Scientific Research: Evolving a National Information Transfer System

T.B. Rajashekar

Scientific growth depends on existing knowledge. Progressive specialisation in science necessitates a researcher to be familiar with what is already known scientifically about the problem to be tackled. In fact, a genuine scientific question cannot be formulated without reference to a background of existing knowledge. A research report is unacceptable, for instance, if it does not cite previous work on the subject. If information that has accumulated with time were no longer available to the scientific community, scientific discoveries and technical innovations would lose ground and probably disappear altogether.

Results of scientific research and technological experience are required not only for the growth of science, but also for the economic, industrial and social progress of a society. The productivity and national wealth of a country depend on technological advances. Technological advancements are possible through innovation and improvement of processes and methods. Innovations are themselves products of scientific research and discoveries.

Scientific growth and societal developments are thus heavily dependent on the ability of the science community to assimilate the results of previous research. But this assimilation is possible only if the results and interpretation of completed research are properly reported and the reports efficiently disseminated throughout the community. Science itself would stagnate if its own achievements were no longer efficiently reported, disseminated, and assimilated in an efficient manner.

Communication and Information Needs of Scientists

Communication and information form integral part of scientific activity. Studies have shown that scientists spend about half their time in communication of one sort or another for giving and receiving information.

There are several reasons why a scientist wants to communicate:

- to exchange information/data needed for scientific work
- to stimulate thought and action through interaction
- to get feedback from peer workers(criticism, refutation, refinement)
- to put claim on one's ideas and findings
- to get recognition (rewards)
- to keep a permanent record for later reference

A scientist requires different types of information at various stages of research. Based on time span covered, one can arrive at two broad categories of information:

1. Background information on the research problem (what has been done so far? What is the state-of-art?). Such RETROSPECTIVE INFORMATION will be specially useful in the beginning stages of research.
2. Current information related to research being carried out (what work is currently being carried out elsewhere?, by whom?, what results have been reported?) CURRENT AWARENESS becomes essential for keeping oneself up-to-date.

Any information provision, retrospective or current, is usually a two-stage process, that of IDENTIFICATION — identification of what information exists which can help to fill the information gap or solve the problem in hand (for example, a list of references with or without abstracts), and ACCESS, getting access to original sources of information (for example, copies of full length papers).

In terms of type of information content, a scientist may need hard data (numeric data resulting from measurement, calculation, observation, experimentation, etc), or soft data (documentary information, ongoing projects, experts, etc.) conventional information transfer methods.

Channels of information transfer within science have been categorised into formal and informal channels which correspond roughly to written and oral communication.

Scientists working at the forefront of their respective fields (often called the 'leaders' or the 'elite' group) tend to form a personal network of information channels (the so-called 'invisible college') for information exchange. This may take the form of correspondence, face-to-face or telephone conversations, preprints, conference, etc. Use of such informal channels ensures rapid information exchange.

But the large majority of the scientific community depends largely on the formal channel — primary and secondary journals, technical reports, patents, dissertations, conference proceedings, etc. Of these the primary journal forms the most important means of formal information transfer.

Problems in conventional information transfer methods

Each link in the 'Information Chain' starting with the information producers (the scholars), editors and publishers of primary journals, conference, etc., professional societies and associations, secondary journal publishers (abstracting and indexing services, distributors and suppliers, libraries and information centres), and the users (scholars) play a very important role in the communication cycle. A breakdown or delays in any of the links in the chain could have serious consequences on the information transfer process.

Conventional information transfer methods (oral and written) are proving to be highly inadequate in the post-industrial, information-rich societies. The single major factor that has contributed to this situation is simply the speed and magnitude of scientific growth — expenditure on science, scientific establishments and scientific workers. It is no longer easy to be aware of all the important workers in a field and to maintain personal contacts. Further, reaching a desired person is not always easy (for instance, it has been observed that three out of four telephone calls find the other person absent, leading to telephone 'tagging'). Given the escalating travel costs and the ever-increasing number of conferences, one can no longer hope to come face-to-face with all the important workers in the field. In very fast changing research fronts (for example, 'Super-conductivity', 'Cold Fusion') communication through correspondence and preprints does not often offer the desired speeds.

The problems are still worse when it comes to formal channels of communication, due mainly to information explosion, literature scatter, publication time lags and publication costs.

Information explosion

With growth in expenditure on scientific research and scientific manpower, there are more research results and practical experiences to be reported. In step with this, the scientific literature is growing at a very rapid pace. This is well exemplified by the growth in the number of published journals in science and technology. Estimates indicate that currently there are close to 50,000 periodicals in science and technology and this number is steadily increasing at a compound rate of 2 to 4 per cent a year. Considering an average of 40 papers per periodical title per year, there are about 2 million journal papers published every year. To this one has to add other publication types like reports, conference proceedings, theses, patents, etc. Overall, scientific literature is estimated to be doubling every 10-14 years.

Besides increase in the number of published journals, there is increase in the number of papers reported in a periodical title and the length of papers, that is, each journal tends to increase in size as more papers are written and submitted for publication. Several are known to be doubling their size every 5 years.

Secondary publications (abstracting and indexing journals) aim to perform the task of scanning and providing synopses of the primary literature. These publications themselves are increasing in size approximately at the same rate as the primary literature. Chemical Abstracts, for instance, took 32 years (1907-38) to publish first million abstracts, 18 years for the 2nd million, 8 years for the 3rd million, 4.75 years for the 4th million, 3.3 years for the 5th million. Currently more than half a million abstracts are being published by CAS every year.

Literature scatter

'Dispersion' or 'Scatter' of the literature is a problem closely related to that of growth. The more a particular sub-field of science grows, the more dispersed the literature is likely to be. In a typical field of research, among a great number of journals, quite a high proportion may actually appear in a relatively small number of 'key' journals in the field. A typical scientist is likely to scan five or six 'key' journals regularly, thereby covering a major portion of papers of his interest. However, he would need to routinely scan very many more journals to increase coverage further.

The problem of current awareness

For those who are not part of the 'invisible college' or the 'elite group' of science, and for whom primary journals are the major means of information transfer, this growth in the volume of literature published and literature scatter, creates great problems in keeping up to date in any field of specialisation. The problem is simply this: the literature of the field grows rapidly, but the time that any individual has to read it, remains more or less the same. A hypothetical scientist spends about 10 per cent of the working day in "keeping up with the literature", and this proportion is the same in 1990 as it was in 1980. Yet, twice as much is published in 1990 as it was published in 1980. Thus, the scientist must either fall further and further behind in current awareness activities or must improve efficiency by using better methods of surveying the literature. This can be illustrated by considering a field as specific as 'Cardiovascular Pathology'. More than 10,000 papers are estimated to be published a year in this field. At an optimistic rate of digesting 4 papers a day, a scientist would take around 10 years to cover this material, by which time another 100,000 papers would have been added.

The only way to keep uptodate effectively is by scanning secondary publications or better yet, by participating in a current awareness service in which a computer is used to search this secondary literature. It is no longer possible to keep well informed simply by scanning a small sample of the primary literature. Even through the use of secondary services scientists are unlikely to discover every paper of potential relevance to their interest, but they might be able to push their coverage up to, say, 90 per cent, which is a great improvement on what one could expect to achieve by scanning only the primary literature.

Publication delay

There are substantial delays in the publication of primary and secondary literature. There may be delays from a few months to more than a year, from the time a paper is submitted for publication to the time it actually appears in print. As more papers are written and submitted for publication, publication "backlogs" develop and greater delays occur because many papers are competing for the limited publication space available. There will also be some delay from the time a research project is completed to the time a paper describing the project is submitted for publication. Majority of science journals are thus more archival than current, reporting research concluded many months ago and perhaps begun years earlier.

Publication costs

The cost of science publications is increasing at a much faster rate than general indicators of inflation in the economy. The publication process is a very expensive one, and publication costs have been increasing extremely rapidly because of increasing costs of labour, materials and equipment. The cost of scientific periodicals is raising by 10-15 percent every year. For instance, the subscription cost of '*Science*' journal went up from $ 117 in 1985 to $ 308 in 1988. The most severe price increases have occurred in secondary publications. Subscription price of '*Chemical Abstracts*', rose from $12 in 1940 to $ 3,500 in 1976, to $ 6,400 in 1984 and to more than $10,000 in 1988.

Majority of the primary journals are beyond the pocket of the individual scientist and are available only in libraries. The more expensive journals are accessible only in the larger libraries. The major secondary services (abstracting journals) are increasingly going out of the reach of the smaller or less wealthy libraries and are available only in larger, wealthier libraries. The general accessibility of the literature declines as a result.

The crisis of libraries

A typical academic library, while doubling its expenditures in less than a decade, finds itself with a budget that buys proportionately less and less of the newly published literature, because the costs of this literature and of personnel to handle it are both increasing much faster than general indicators of inflation in the economy. It is not only 'information' in the form of published literature which is becoming expensive, but also efforts to make this information accessible (acquiring, processing, storing). At the same time the literature continues its inexorable growth, and many libraries, despite being unable to "keep up" with this growth, face acute shortage of space.

The solution: electronic information handling

The solution to the problems in conventional information transfer has come mainly as a result of a major shift in information handling — from information in its analog and physical form (voice, print, audio visual) to electronic form. The modern information situation is a direct consequence of the digitisation of information, opening out a new paradigm in information handling. Electronic information offers enormous speed and convenience for creation, processing, transmission, storage, retrieval, and presentation over oral and written forms.

It is the application of information technology that has helped in the conversion of more and more information into electronic form. Specifically, it is the rapid developments in key information technologies like computer processing, data storage, data transmission and software that have brought about this transformation.

We are already witnessing several products and developments which have vastly improved the science communication scenario, affecting almost every link in the 'Information Chain' mentioned earlier. Some of the major developments of the international scene are discussed below.

Information databases and giant library catalogues

Currently, more than 4,000 databases (full text, numeric - scientific/commercial, reference) are available on database host computer systems which can be accessed online through telecommunication networks. DIALOG, BRS, ORBIT, PERGAMON/INFOLINE,ESA/IRS, STN International are some of the well-known database hosts providing online information retrieval. More than 300 million items of information are estimated to be available in these databases — ranging from references to literature, to the complete text of a journal article; to a directory type listing of companies, associations or famous peoplc; to an in-depth financial statement for a particular company, etc. Giant library catalogues (for example, Library of Congress, British Library, OCLC — Online Computer Library Centre, WLN — Western Library Network, etc.), containing several million library catalogue records along with the library location codes, are available for computer search and retrieval.

Several of these databases and catalogues are available on 5 1/4" CD-ROM discs which are portable and compatible with micro-computers.

A 650 MB CD-ROM disc can store more than 2,50,000 pages of information or the entire

En-cyclopaedia Britanica, with some more space left. A personal computer with a CD-ROM drive comes closest to the concept of a desk top global information system. Eraseable optical storage systems, compatible with micro-computers, are expected to further simplify and localise database creation efforts.

Electronic document — preparation, distribution and delivery system

Document imaging systems like scanners and character recognition systems like OCR (Optical Character Recognition), graphics systems, text processing systems (for example, 'TEX' by Donald Knuth), delivery systems like telefax, floppy diskette and CD-ROM, combined with high speed data transmission systems like packet switching, are helping to realise electronic publication systems (e.g., 'Electronic journal') and paper less information systems. Starting with preparation of papers for publication using word processing systems, the entire publication process can be in electronic form, including submission of manuscript in electronic form, refereeing, editing, etc. From this stage, the 'electronic paper' can be used either for producing a printed publication (as part of a journal) or become part of a full text electronic database.

So as to improve the efficiency of their operations, world's leading document supply centres like BLDSC (British Library Document Supply Centre) and National Library of Australia entertain document supply requests in the form of electronic mail messages. Several studies have been carried out, including the ADONIS project at BLDSC, for evaluating the use of CD-ROM based full text documents (journals) for efficient search, retrieval and supply of documents.

Electronic mail and measuring systems (EMMS)

Computer networks, particularly academic and research networks like ARPANET (Advanced Research Project Agency Network, USA) and JANET (Joint Academic Network, England) and cooperative networks like BITNET, USENET have significantly improved the speed and convenience of communication among researchers. E-mail (Electronic mail) is the most widespread service and use of such networks. It allows a user to send a message to another user on either the same or a different host. Mail is typically a one-to-one, point-to-point communication medium, though it is possible to mail to more than one recipient at the same time. For the same message, E-mail is cheaper than trunk dialed telephone calls due to more efficient use of communication lines. Furthermore, E-mail dose not assume or require the presence of the concerned party at the other end. File transfer is the second most common service. It allows files to be transferred from one host machine to another. Remote-login is another important service. It allows a user to access a system on a remote host over a network as if connected to a direct terminal. Computer conferencing systems are message exchange systems that are generally similar to electronic mail systems, but differing in that they are many-to-many (or broadcast) rather than one-to-one media. Usually one copy of a message is kept per host rather than one per user as for mail. Bulletin board system allows users to post news into and receive news from a central database (bulletin board).

Library networks

Libraries, besides automating their operations, are participating in computer network, for sharing information resources (for ILL — inter-library loan request and document transmission through E-mail, telefax), information access efforts (cataloguing), for online information retrieval, etc. Majority of libraries in developed countries have been computerised (particularly library catalogues). Communication infrastructure provided by data network like TYMNET, TELENET, EURONET, etc., has been taken advantage of for implementing library networks in U.S.A. and Europe. Heterogeneous systems are being interconnected using OSI (Open System Interconnection) model of ISO.

The scholarly workstation

Scientists need computers for data analysis. They need to communicate (exchange information) with other scientists, search databases for information retrieval and create and edit documents. It is possible to integrate all these needs of a scientist into a 'Scholarly workstation'. Such a workstation can be realised in the form of a micro-computer which can be used for several purposes. As a stand-alone system, it can be used for computation, processing, information creation (text processing) and local information storage. Through a suitable communication link, it can become a node in a LAN (Local Area Network) or WAN (Wide Area Network) or simply a terminal to an external computer system, and used for purposes of message and file transfer, for information search and retrieval from remote databases, for using an external computer system through remote log-on computer conferencing, etc.

The Indian scenario

A researcher in a developed country is at a far greater advantage than his counterpart in India due to the ease with which he can get access to information through computer databases and exchange information with his peers using academic and research networks. Quick access to information and information exchange among researchers have meant improved quality of R&D projects at reduced time durations and cost, better utilisation of information resources and sharing of information access efforts.

Since 1988, a major effort has been made in India for improving the communication among the academic and research community. An academic and research network is currently evolving in the country, under the ERNET (Education and Research Network) project of DOE (Department of Electronics). An interim WAN (Wide Area Network) has been set up connecting eight leading academic research institutions in the country (five Indian Institutes of Technology, Indian Institute of Science, National Centre for Software Technology, and ERNET office in DOE). The interim WAN consists of a central E-mail facility (a VAX computer available 24 hours a day) located at NCST and forming the hub of a star network. It can be reached by a dial-up line from other ERNET sites. This node is connected to academic and research networks like BITNET, USENET in the USA via UUNET in Virginia through a dedicated line, and to similar networks in Europe, via MCVAX gateway in Amsterdam. Some 1,000 messages, including 500 for abroad, are handled daily on the NCST central node.

The message traffic has been growing at the rate of 10 per cent per month. The network has significantly helped the scientists in these institutions write joint papers with foreign researchers.

The full capacity of this academic network will be realised when the WAN is implemented through satellite channels. ERNET services include electronic mail, file transfer, remote log-on (terminal access) and conferencing, all of great interest to the academic community. Electronic mail is expected to constitute 50 percent of the traffic on this network, when fully implemented.

Since the introduction of micro-computers on a large scale in India in the early 1980's, efforts have been made in evolving electronic information systems for research and development. The national information centres operating under NISSAT (National Information System for Science and Technology), DSIR and UGC, defence science (DESIDOC), MEDLINE information services provided by NIC-ICMR project, information centres operated by several research institutions and government departments, etc., are providing computer-based information services to the S&T community. Online access to international database hosts has been made easier with the setting up of packet switch gateway nodes in the country. As far as library automation is concerned, libraries are being automated mainly at sub-systems level. Several training programmes in information technology applications for information handling have been launched. About a dozen indigenously developed software packages for library automation are now available in the Indian market. CD-ROM databases have been acquired by a few institutions for providing on-site information services.

Database creation efforts are slowly increasing (for example, the National Union Catalogue of Scientific Serials in India — NUCSSI of INSDOC, Machine Tool Abstracts database of CMTI, Bangalore, Food Science database of CFTRI, Mysore). In the private sector, Informatics India, Bangalore, has played a significant role in provision of database and information services.

INFLIBNET: A plan for national information transfer system in India

Thus, the situation in India now is ripe for integrating these diversified efforts and evolve a mechanism for improving the information transfer situation in the country. Such a mechanism has to take into cognisance several important requirements:

- modernisation of library and information facilities, services and information handling practices using appropriate information technologies;
- inter-connection of such facilities for resource sharing and scientific communication,
- provision of gateways to access information and people abroad;
- improvement of skills of library and information personnel;
- improvement of document access capabilities and services; and
- development of databases for internally generated information and relevant international level databases.

Significant developments in this direction have taken place in the past couple of years for interlinking (networking) of library and information centres in the country with a view to enhancing the information access capabilities of the users from local to regional and national levels. Several proposals have been formulated for establishing library and information networks — INFLIBNET (INFormation and LIBrary NETwork) which is a programme for modernisation and nationwide networking of university and R&D libraries and information centres, DELNET and CALIBNET — plans for networking of libraries and information centres in Delhi and Calcutta. A nationally networked technology information system (TIFACLINE) is also being set up by the Technology Information, Forecasting and Assessment Council (TIFAC). TIFACLINE will primarily aim to integrate technology information available in various government departments, institutions, laboratories and R&D agencies in the country.

INFLIBNET is a project proposal for establishment and development of a nationwide computer communication network linking libraries and the academic and research community, during the Eighth Five Year Plan (1990-95). The motivations for formulating a plan of the magnitude of INFLIBNET are quite obvious. The country is incurring huge expenditure in foreign exchange for procuring library books and journals, which is estimated to be about Rs.500 crores a year by all libraries. The cost of books and journals is rising by 10-15 percent each year. The library budgets are increasing by the same margin, although may not be in real terms. There is substantial duplication in library holdings (for example, about 110 copies of '*Chemical Abstracts*' are being procured at a cost of nearly Rs. 3 crore a year). One has also to consider the substantial duplication in technical processing (classification, cataloguing) of common books acquired across libraries.

In spite of this huge expenditure, a user is handicapped by lack of library resources. It is neither feasible nor possible to fund all the libraries to be self-sufficient in their collection.

INFLIBNET plan was prepared in 1988 at the initiative of University Grants Commission, by an inter-agency working group consisting of members drawn from different government agencies, having expertise in library, information, computers, communication and networking. The working group undertook a detailed study of national and international scenario, conducted an in-depth survey of the status of information technology applications to libraries in universities and R&D institutions and held consultations with librarians and academics, while preparing the INFLIBNET proposal. The report, finalised in December 1988, comprises 12 chapters dealing with aspects like introduction, INFLIBNET organisation, services, standardisation, software requirements, hardware specifications, communication architecture, data capturing and preparation, manpower, management mechanism, implementation mechanism and cost implications. As many as 15 appendices give valuable information supporting the main text. A detailed analysis of the 129 questionnaires returned by university libraries provides a glimpse of the status of university libraries in the context of information technology applications.

INFLIBNET objectives

The main objectives of INFLIBNET are:

a. To evolve a national network, interconnecting various libraries and information centres in the country and to improve capability in information handling and services.

b. To provide reliable access to document collection of libraries by creating online union catalogue of monographs, serials and non-book materials (manuscripts, audio-visuals, computer media, etc.) in various libraries in India.

c. To provide better access to worthwhile bibliographic information sources, with citations and abstracts, such as periodical articles, conference papers, preprints, technical reports, standards and specifications, patents, monographs, etc., through indigenously created databases of the Sectoral Information Centres of NISSAT and UGC Information Centres and such others and by establishing gateways for online accessing of international databases held by international information network and centres.

d. To provide document delivery service by establishing document resource centres around libraries having a rich collection of documents.

e. To optimise information resources utilisation through shared cataloguing, inter-library loan service, catalogue production, collection development and avoiding duplication in acquisition to the extent possible.

f. To implement computerisation of operations and services in the libraries and information centres of the country, following a uniform standard.

g. To facilitate scientific communication amongst scientists, engineers, researchers, social scientists, academics, faculties and students through electronic mail, bulletin board, file transfer, computer/audio/video conferencing, etc.

h. To enable the users dispersed all over the country, irrespective of location and distance, to have access to information regarding books, monographs, serials and non-book materials by locating the sources wherefrom available and to obtain it through the facilities or new communication technologies and union catalogue of documents.
i. To create database of projects, institutions and specialists for providing online information service.
j. To encourage co-operation among libraries, documentation centres and information centres in the country, so that the resources can be pooled for the benefit of helping the weaker resource centres by stronger ones.
k. To develop suitable professional manpower of appropriate quality to establish, manage and sustain the INFLIBNET.
l. To evolve standards and uniform guidelines in techniques, methods, procedures, hardwares and softwares services and so on and promote adoption in actual practice by all libraries, in order to facilitate pooling, sharing and exchanging resources and facilities towards optimisation.

INFLIBNET is a major national effort to improve information access and transfer capabilities that support scholarship, learning research and academic pursuits. By networking, individual libraries will be spending less than if they were to operate independently due to increased productivity.

INFLIBNET services

INFLIBNET is a multiple function/service network. It will provide the following services:

- Catalogue based services:

- Shared cataloguing
  (Sharing of information access efforts)
- Union catalogue
  (Shift from 'Local' to 'National' holdings)
- Online catalogue access
  (What is available, where ?)
- Catalogue production
  (Multiple access media)

• Database services:
- Bibliographic database services
- Retrospective, S.D.I., Current Awareness Services
- Gateway access to Information Data bases/Networks
- Non-Bibliographic databases

• Referral & Directory based services
- Projects (Ongoing/Completed projects)
- Specialists
- Institutions

• Document supply services
- Inter-library loan request processing
- Document delivery (FAX/Non-FAX)

• Collection development:
- Assist selection and procurement
  (Based on Usage/Demand statistics)

INFLIBNET functioning and organisation

INFLIBNET operates at different levels — national, regional, sectoral and local. The national centre will serve as an apex body, managing, overseeing and coordinating the affairs of INFLIBNET. No databases will be maintained by the national centre. There will be four regional centres — one each in north, south, east and west, maintaining union catalogue of library holdings of the region (books, journals). Regional centres will also maintain non-bibliographic databases like projects, experts and institutions. The catalogue of libraries will be aggregated bottom-upwards, from college/department to university library and to regional centre.

At the sectoral level, UGC information centres and NISSAT sectoral information centres and other centres which perform national level functions will acquire, create and access retrospective and current bibliographic databases and offer a variety of modern information services.

As many as 100 universities and R & D institutional libraries will be designated and equipped to serve as document resource centres for purpose of document delivery services.

End users will be served mainly at local levels, such as college, department, university and R & D institution.

Fig. I provides a schematic view of INFLIBNET organisation.

Conclusion

If current developments in information systems technologies are any indication, we are fast approaching the realisation of Vannevar Bush's 'Memex' system. Very large knowledge databases characterise such a system. Four key technologies and their integration will pave way for creation of such a system — data conversion systems for automatic database building, high volume optical data storage systems at affordable prices, high speed data transmission systems using packet switching with all digital communication systems (ISDNs capable of handling image, sound and data over the same lines) and searchware capable of handling natural language

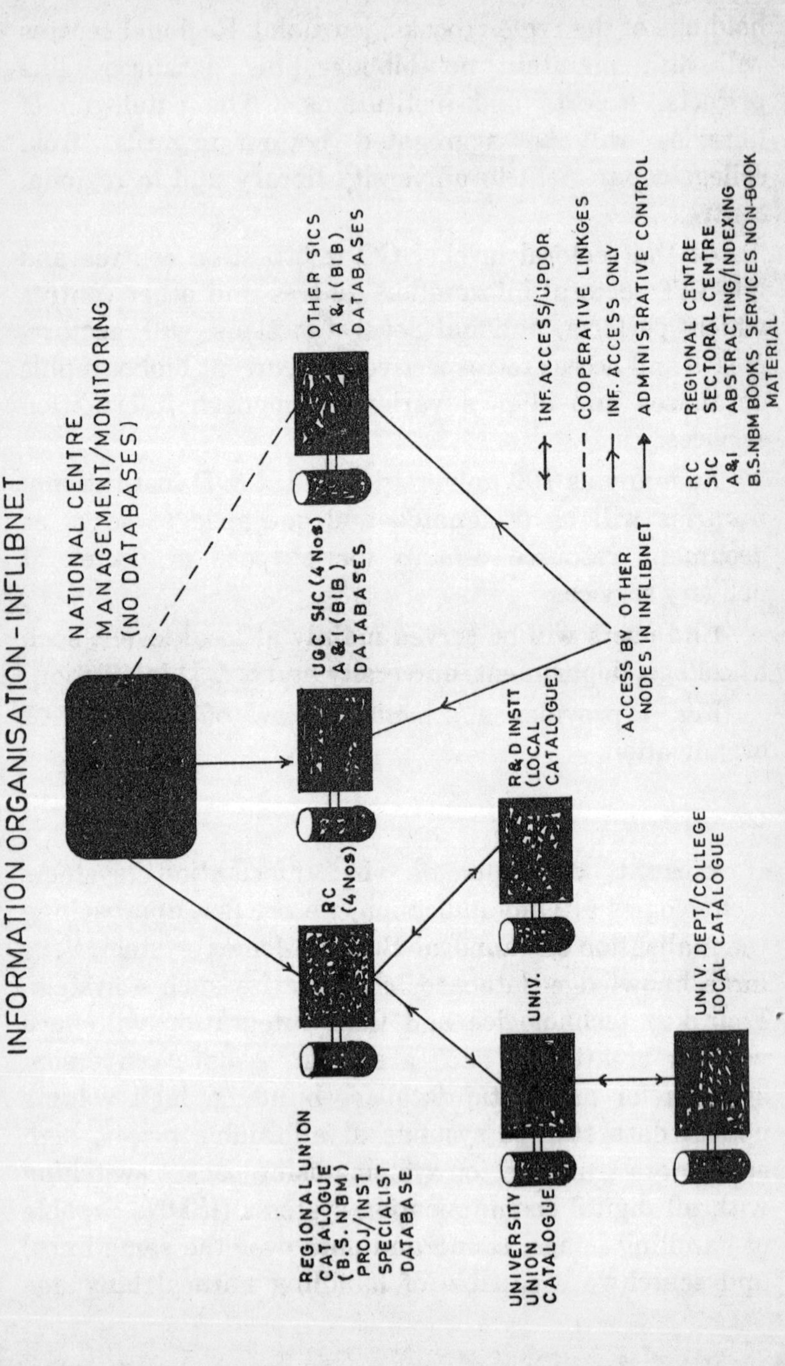
INFORMATION ORGANISATION - INFLIBNET
NATIONAL CENTRE
MANAGEMENT/MONITORING
(NO DATABASES)
RC
(4 Nos)
UGC SIC (4 Nos)
A&I (BIB)
DATABASES
OTHER SICS
A&I (BIB)
DATABASES
REGIONAL UNION
CATALOGUE
(B.S. NBM)
PROJ/INST
SPECIALIST
DATABASE
R&D INSTT
(LOCAL
CATALOGUE)
ACCESS BY OTHER
NODES IN INFLIBNET
UNIV.
UNIVERSITY
UNION
CATALOGUE
UNIV. DEPT/COLLEGE
LOCAL CATALOGUE
INF. ACCESS/UPDOR
COOPERATIVE LINKGES
INF. ACCESS ONLY
ADMINISTRATIVE CONTROL
RC REGIONAL CENTRE
SIC SECTORAL CENTRE
A&I ABSTRACTING/INDEXING
B.S.NBM BOOKS SERVICES NON-BOOK MATERIAL

FIG.-1

and providing high precision context based information retrieval with improved screen presentation (using AI techniques and parallel processors).

It is visualised that a global information system will emerge in the next fifty years, wherein AI interfaces and gateways, communicating across digital networks, will link the innumerable information systems worldwide. In such a scenario, libraries will depend more on networking with shared databases across distributed systems. Local information files will be electronically linked to regional and national databases. Libraries will then become "Knowledge servers" and librarians will be search/database specialists.

If we desire to be part of such a scenario, we should start moving, right now, in that direction. Ideas and plans such as that of INFLIBNET provide an opportunity and means towards that goal.

References

1. T.M. Aitchison, The Database Producer in the Information Chain. *J. Inf.Sci.* 14,1988, pp.319-327.
2. Simon Bell, Information Systems Planning and Operation in less Developed Countries. Part l: Planning and Operational Concerns, *J. Inf. Sci.* 12, 1986, pp.231-245.
3. Simon Bell, Information System Planning and Operation in less Developed Countries. Parl 2: Case study, Information System, Evaluation. *J. Inf. Sci.* 12,1986, pp.319-331.
4. Alan Burns, (ed), *New Information Technology*. Chichester: Ellis Horwood, 1984.
5. C.W. Hanson, *Introduction to Science Information Work*, ASLIB, 1971.

6. Allen Kent, and Thomas J. Galvin, (eds), *Information Technology: Critical Choices for Library Decision-makers*, New York: Marcel Dekker, 1982.
7. Douglas A.Kranch, The Development and Impact of a Global Information System, *Inf.Tech.Libs.*, December 1989.
8. F.W.Lancaster, *Toward Paperless Information System*, New York: Academic Press, 1978.
9. F.W.Lancaster, *Whither Libraries? or, Whither Libraries*, Coll.Res.Libs. July 1989 (A C&RL classic reprint).
10. M.N.Mathur, and S.Ramakrishnan, *Project ERNET: Perspectives, Plan and Approach of an Academic and Research Network Program in India*, In: CCDC-87, ICCC Special Conference, 1987.
11. A.J.Medows, *Communication in Science.*, London: Butterworths, 1974.
12. Office of Technology Assessment, U.S.Congress, *Information Technology Research and Development: Critical Trends and Issues,* New York: Pergamon, 1985.
13. John S. Quarterman, and J.C. Hoskins, Notable Computer Network, *Comm.* ACM.29,1986, pp.932-971.
14. University Grants Commission, Govt. of India, *Development of an Information and Library Network: Report of the inter-Agency Working Group*. New Delhi, U.G.C., 1988.
15. Paul Wasserman, Invention and Design of Information Products and Services, *Infomediary, 1985, pp. 11-16.*
16. John Ziman, *Knowing Everything about Nothing: Specialisation and Change in Research Careers.* Cambridge: Cambridge University Press, 1987.

17. John Ziman, *The Force of Knowledge: The Scientific Dimension of Society*, Cambridge: Cambridge University Press, 1986.
18. Peter Zorkoczy, *Information Technology: An Introduction,* London: Pitman, 1982.

Science, Peace and Development

Rais Ahmed

There was a time when it was believed by a vast majority of scientists that science had nothing to do with larger socio-economic and political issues, and their interactive relation to science problems. Then, there was a belief among some scientists that science *per se* could bring about social progress. The symbiotic relation between Science and Society is being gradually appreciated. The interconnections between science, research and innovation on the one hand and non-alignment, self-reliance and dilemmas of development are now being examined on the other. Even global issues like technology as a tool of domination of a small number of countries over the rest of the world, restrictions on scientific information and disputes about patent laws are now a part of the scientists' concern. Of course, the arms race, the nuclear weapons, and the great defence industries which are creating a dependence structure of their own, and questions of world peace and regional peace are matters for the anxious consideration of scientists.

These complex issues are being appreciated, but not quite understood, as scientists would like to deal with more explicit formulations in regard to these questions. But this will come about only if the scientists take deep interest in these problems, analyse them with their typical objectivity, and move on to action in this sphere. We believe that at this stage of India's developmental history, mere discussion or analysis is not an adequate response; without action, desirable changes cannot be brought about. At this point inhibition towards social action must be given up by scientists. This can create a healthier atmosphere for social and scientific development, and for the solutions of relevant social and economic problems. The Forum of Scientists for Peace and Development in which many leading scientists participate, is one such forum for scientists to come together, think and act for real progress.

Even a cursory examination of issues of science, peace and development reveals an immense complexity of problems worthy of the attention of scientists. We can cite just a few relevant observations. About 1.2 percent of the GNP is being spent in our country on science and technology and expenditure has gone from about Rs 500 crores to Rs 3,000 crores, which is a factor of 60 in about 40 years. The picture is less rosy when we remember that a rupee now is worth less than 10 paisa of 1950; and the number of scientists working in various government related organizations has certainly increased by a factor of more than six. This implies that perhaps less money is now available in our country per working scientist. Especially so when today, the cost of research is approximately the same everywhere and costs have gone up with greater sophistication of the instrumentation of research. Again, there is a question of the distribution of this money as between agencies and universities, and as between defence related research, development of weapons and delivery system, etc., and research on health and nutrition, spread of education and on the nature of development itself. There are also questions related to the organization of research, that is institutional functioning, linkages amongst institutions, utilization of research, and the value system operating among research scientists. These questions are related to whether we are making optimum use of whatever financial and human resources we have available in our country. Is science contributing to the raising of GNP itself? A UNESCO study on 1200 research centres conducted in the late '70s has shown a high correlation between non-financial and non-academic factors, which can be generally expressed by the term Ecology (such as human conduct, integrity, objectivity, fair play, etc.) of science. We have here a number of highly relevant questions which should be discussed by scientists and conclusions must be reached about any alteration of policy if that is called for.

Development, as we all know, can no more be gauged by GNP or per capita income. Even industrialization by itself does not mean development, and much less, availability of all sorts of consumption durables in the shops. The concern of development must inevitably be to improve the conditions of living of people as whole — that is what Jawaharlal Nehru's 'Tryst with Destiny' was all about, and the reason for the whole freedom struggle. Because, if vast sections of population live without basic necessities of life (under-nourishment, ignorance, insanitary conditions of living and inadequate protection of health, unemployment, etc.), they will only be driven to shake up the society. Social justice and democracy have, therefore, been accepted as basic tenets of our Republic. The prevailing conditions in our country in regard to factors mentioned above are such that normally more than half of the population survives in various degrees of deprivation, contributing to frustration, discontent, and sometimes violent outburst. The inner strength of the nation, which is one of the greater factors in national security, is being sapped.

There are many other questions of far-reaching significance which the scientists must examine in depth — the degradation of environment in which the poorest including the tribal people suffer the most, the impact of such an important social factor as the green revolution on the poor farmers and the landless, the problem of generating power from nuclear sources, the tremendous debt burden and the balance of payments both of which adversely affect science and education. Perhaps, a diversion of resources is called for in order to progress in the desired direction, and in order to give a fillip to scientific work, and adventure of ideas. Perhaps, reconstructing of administration, management and social relations has to find a place on our agenda today. A new approach to planning with a different set of priorities has already been talked about and we have to see that it does not remain a mere play of words.

This very brief and inadequate statement shows, at least, that we have most complex problems in our lap, which require peace in the first place — not threats of armed struggle or regional wars which drain our energy and resources. Our heritage from Buddha to Mahatma Gandhi also gives us a natural inclination to look for peaceful solutions to personal as well as social and inter-state problems. It was Mahatma Gandhi who unambiguously raised his voice against the atomic bomb, and Jawaharlal Nehru who in 1954 gave the first call for banning of nuclear tests. It was a remarkable event when in 1986 Mikhail Gorbachev, President of USSR and Rajiv Gandhi, the Prime Minister of India, gave a call for a non-nuclear weapons and non-violent world. The world has drawn back from the precipice partly because of these events and because of world public opinion which wants confrontation to be replaced by dialogue and consultation. Far-reaching changes in Africa, Asia and Europe have been possible because of this. Resources for peaceful reconstruction of economics are being released — though very slowly and with great reluctance on the part of those directly involved in the armament establishments.

Nearer home, the issues are bound to be clouded by emotions and overtones of national pride. In full knowledge of these factors, scientists in India have firmly expressed themselves in favor of world peace and peace in this region. We stand squarely for the unity, integrity and security of our country, and we know that national security is multi-dimensional; it is based on the cohesive strength and economic health of our people as much as it is based on the defence of our borders. We, therefore, plead to the scientific community to firmly stand for peaceful and consultative solution of problems, however complex and inflammatory they may happen to be. We further plead for greater participation of scientists, through research, scientific analysis of social problems and education and many forms of public activity to ensure a people oriented and self-reliant development of our economy.

Science, Society, Government and Politics — Some Remarks on the Ideas of Jawaharlal Nehru

Obaid Siddiqi

Nehru was passionately interested in larger ideas, in history, culture, progress and the destiny of mankind. In this he was different from the other leaders of the Indian National Congress who were preoccupied with constitutional issues and were generally conservative. Mahatma Gandhi was an exception, but his concerns, apart from politics, were morality and religion.

Address to Science Congress

Between 1938, when Nehru first sent a message to the Science Congress at Calcutta, and 1963, when he last spoke to the Science Congress at Delhi, Nehru addressed this gathering of Indian scientists sixteen times, expressing solidarity with Indian science and exhorting scientists to play their part in the destiny of this nation. Baldev Singh, Secretary of the *Zaheer Science Foundation*, has done yeoman's service by collecting the speeches and writings of Nehru on Science and on Science and Society in two volumes published by the Nehru Memorial Museum and Library.

Nehru's exhortations to scientists centre around three principal themes. First, science must serve the people — it must raise their standard of living. Second, the scientific way of thought, the scientific temper, must spread to all spheres including politics and third, science must be tempered with wisdom, it must pay attention to the spiritual side of man. In his later years, as Nehru became increasingly aware of the threat of nuclear war and the destructive powers of science, he spoke out against militarism and for peace and internationalism.

In the next few paragraphs, I propose to recall some of Nehru's beliefs and raise the question, how far and how well his ideas have stood the test of time and to make some remarks on the present state of our science and society — the state of the "Nehru legacy" so to say.

Nehru's world view

Nehru's mind did not work in rigid frames of thought. He was open- minded to a fault, always willing to look at the two sides of a question, often saying different things at the same time. It is, therefore, very easy to blame him for inconsistency, as some of his critics do. This is unjust. While he subscribed to the fundamentals of socialism, its "numerous inner controversies" did not interest him. He did not like dogmas and was irritated by the fact that "the rigidity of dogma that was chiefly in religion was projecting itself in the realm of politics and economics and invading the scientific sphere".

Nehru's radical outlook changed somewhat with age and rather inevitably after assuming power over the State. He remarks, in *Discovery of India*:

> "My early approach to life's problems had been more or less scientific, with something of the easy optimism of the science of the nineteenth and early twentieth century. A secure and comfortable existence, and the energy and self-confidence I possessed, increased that feeling of optimism. A kind of vague humanism appealed to me."
>
> "Those days of my youth seem far away now, not merely because of the passage of years but far more so because of the ocean of experience and painful thought that separates them from today".

This frank openness to change is a particularly endearing feature of Nehru's personality.

Nationalism

Nehru was proud of his racial ancestry and India's past intellectual attainments. He took great pride in the fact that the people of Asia — Indians, Egyptians and Chinese were capable of subtle thought when the rest of the world had not seen the dawn of civilization. But his idea of India and Indians was not narrow. It included everything that has happened to this vast subcontinent over its long history. Speaking at Science Congress at Hyderabad in 1954, Nehru remarked, "And so, after this millennia of history, we are a peculiar mixture, carrying the burden of the wisdom, the folly and the madness of 10,000 years; and it is all mixed up and sometimes the wisdom comes up, often the folly, and sometimes the madness. What are we to make all of this? How does science help us to bring a modern synthesis of these problems — something in keeping with the genius of

India? Because it has been the genius of India to work for synthesis, in spite of all those who have tried in the past to break it up, in spite of all those small-minded persons without vision, who today think parochially in narrow terms, forgetting the larger issues before this country or the world".

Nehru's nationalism was not bigoted. Indeed, he was an internationalist. In the *Discovery of India,* he states: "What is my inheritance? To what am I an heir? To all that humanity has achieved during tens of thousands of years, to all that it has thought and felt and suffered and taken pleasure in, to its cries of triumphs and its bitter agonies of defeat, to that astonishing adventure of men which began so long ago and yet continues and beckons to us. To all this and more, in common with all men."

He warned the scientists repeatedly against succumbing to nationalism and nationalistic pressures from their governments and countrymen.

It is this philosophic outlook which is reflected in the well- known sentence of the scientific policy resolution, "It is an inherent obligation of a great country like India with its traditions of scholarship and original thinking and its great cultural heritage, to participate fully in the march of science which is probably mankind's greatest enterprise today." One cannot but note the contrast between this spirit and the feeble assertion that a poor country cannot afford the luxury of pure research.

Religion

Nehru was not particularly religious — not what he called a "religious type of man". He described Vinoba Bhave as a "religious type" of man and called himself a pagan because like a pagan he loved the pleasures of life. He remarked that "being a pagan, my mind is open to all ideas. It does not reject an idea unless after an examination I do not find it good enough." (Science Congress, Madras, 1958).

Religion, as practised by its followers seemed to Nehru, to be closely associated with superstitious, beliefs and dogmatic practices, "There was an element of magic about it, an uncritical credulousness, a reliance on the supernatural". He advocated a philosophic view specially "for men at the top", but was suspicious of mysticism and metaphysics. He remarks, in *Discovery of India*:

> "There have been great mystics, attractive figures, who cannot be easily disposed of as self-deluded fools. Yet mysticism (in the narrow sense of the word) irritates me; it appears to be vague and soft and flabby, not a rigorous discipline of the mind but a surrender of mental faculties and a living in a sea of emotional experience. The experience may lead occasionally to some insight into inner and less obvious processes, but it is also likely to lead to self-delusion." And "always we must hold to our anchor of precise objective knowledge tested by reason and even more so by experiment and practice, and always we must be aware of losing ourselves in a sea of speculation unconnected with day-to-day problems of life and the needs of men and women. A living philosophy must answer the problems of today".

Public enterprise

Nehru was a passionate advocate of public enterprise and somewhat suspicious, perhaps unfairly suspicious, of lawyers and businessmen.

"In India, we were dominated by a lawyer's mentality and more lately by a classical philosopher's. Latest in the picture was the businessman's mentality. But neither the lawyer nor the classical philosopher can solve our problems. Present-day problems of India are to be tackled not with the approach of a lawyer or classical philosopher but with a partly scientific and partly engineering approach. I doubt very much if the businessman with his limited outlook can play any part at all." (Science Congress, Poona, 1950).

"Private enterprise will not plan for the growth of a nation, but will plan for the growth of its own enterprise. Surely the issues are too grave to be left to chance or to the motives of the private enterprise person, however evil or good he may be.... That was 19th century economics." (Ceylon Association for Advancement of Science, Colombo, 1962).

And yet one of the most innovative and bold contributions that Nehru made to economic philosophy and planning is the idea of mixed economy. As was to be expected, neither the right, nor the left is happy with this policy, but neither knows how to get away from it. Whatever be the shortcomings of Nehru's philosophy of mixed economy, it seems this philosophy is going to prevail in both socialist and capitalist systems.

The role of science

Let me now turn to Nehru's view of science and his science policy. At Cambridge, Nehru read for the Natural Sciences Tripos, and was influenced by the ideas of Darwin, Karl Marx, Bernard Shaw and Fabian socialists. He was moved by the Russian revolution and impressed by Lenin's determination to build a society based on egalitarian and scientific principles. After returning to India in the twenties, prolonged incarcerations gave Nehru time to read and think more deeply about science and society. This period of Nehru's mental evaluation is well reflected in *'Discovery of India'* and *'Glimpses of World History'*. A perceptive recent account of Nehru's intellectual development appears in Baldev Singh's article in the volume on Nehru's writings on Science and Society.

In the Congress party Nehru attempted to evolve a policy on science. In this he was ably assisted by Professor Meghnad Saha. Nehru's belief in the social function of science had been greatly reinforced by the large scale participation by scientists in England and America in the war efforts. After he became the Prime Minister of India, Nehru earnestly set about reconstructing science in India and putting it to use.

Nehru's view of science was frankly Baconian. His strong interest in science arose in his own words from "the social consequences of science than science itself" (Science Congress, Bombay, 1960). Nevertheless, this interest was very strong and science had a high place in his scheme of things as it is evident from his following remarks:

> "Science is the spirit of the age and the dominating factor of the modern world. Even more than the present, the future belongs to science and to those who make friends with science."

"It was science alone that could solve these problems of hunger and poverty, of insanitation and illiteracy, of superstition and deadening custom and tradition, of vast resources running to waste of a rich country inhabited by starving people. (Message to Silver Jubilee Session, Science Congress, Calcutta 1938).

His strong interest in science becomes obvious from his remarks on various other occasions, for example:

> "I am convinced that of all the big problems that face India today, nothing is more important than the development of scientific research, both pure and applied and scientific method." (Message to Silver Jubilee Celebrations, Royal Institute of Science, Bombay, 1945).
>
> "You cannot divorce science from social and political happenings and from the economic structure of the world. Therefore, perhaps it is time that science developed a certain philosophy and unity; if I may put it so, it had this quality in the olden days when science presented a small picture than it does now." (Science Congress, Delhi, 1947).

"There is no other way except the way of science ultimately for the development of human life and institutions. This is the scientific approach to life's problems. I myself doubt very much if even scientists have the scientific approach to life's problems. They are often unscientific outside their laboratory or in considering other human problems.... The fundamental thing is a scientific approach."(National Institute of Science, New Delhi, 1948).

Nehru did not distinguish much between pure science and applied science or between science and technology. To him they were all "about the same". He was not against pure research which is "essential and out of which come many practical applications" but admitted that a person like him "concerned with day to day problems of great importance has always to think a little less of pure research and more of application of research to problems of human society." But above all science to him was human activity and no human activity could be worth anything unless it is carried out with the larger humanity in view.

"Whether it is pure science or applied science, it must serve humanity, human good, raise human standards and help our fellow men." (Central Electrochemical Research Institute, Karaikudi, 1948).

Nehru's vision of science is embodied in the Scientific Policy Resolution adopted by the Government of India in 1958, an extremely well written and inspired document. I should add that to implement his policy, Nehru found uncommonly able assistants in P.C. Mahalanobis, S.S. Bhatnagar and H.J. Bhabha who applied their considerable talents to translate Nehru's vision into practice in the sphere of economic planning and scientific research. They helped Nehru weave his economic and technological inclinations into a coherent policy of planned growth. India thus began the rapid expansion of its scientific and industrial infrastructure on a scale which had not been dreamt of before.

Not all scientists were happy with Jawaharlal Nehru's policies in post-Independence India. Meghnad Saha now sat in the Parliament with the opposition. Sir C.V. Raman was highly critical of the policy of big science in big laboratories. "The essence of science" he maintained "is independent thinking and hard work and not equipment."

The Nehru legacy

I will now turn to the outcome of Nehru's science policy which has been followed more or less unchanged by his success over these past 25 years. I intend to make a few remarks about the success and failure of this policy, but I should add that it is very difficult to be objective and detached in this assessment. The issues are far too complex and we are too close to the events, in a sense too much in them.

Expansion of science and technology

There is no question that the pursuit of Nehru's policy has led to a remarkably rapid growth and expansion of scientific and technological research in India. One need not cite numbers of men and women engaged in science or the numbers of research institutes and laboratories to prove this point. It is enough to look at the proportion of GNP spent on scientific research. In 1947, the figure was somewhere around a tenth of 1 per cent. Currently it stands at 1.1 percent. Ministries, scientific agencies and science departments constitute a significant portion of governmental structure. India stands apart from the other third world countries, perhaps with the exception of China, in that science and technology constitute a significant part of its national developmental efforts.

India has industrialised rapidly and increased its agricultural production very considerably. The extent to which indigenous research has contributed to growth is a point which needs more careful examination. There is no doubt that in certain sectors of industry and in agriculture, the contribution of indigenous research to production has been substantial. In other sectors of the economy it could have been more.

At any rate, thanks to Nehru's policies, India has acquired a large reservoir of manpower and technical infrastructure which should serve it well in the time to come and allow it to become self-reliant and strong.

Failures

While there is little doubt that Nehru's policy has served science well, has science done its bit for society? The belief is widespread that the application of science to fulfil our needs and to improve our lives has lagged. In public perception "science has not delivered the goods." One consequence of this feeling is an inclination among many scientists to search for relevant science. This urge is strongly reflected by movements such as the People's Science Movement.

Although science has expanded, is it good enough? Is its quality commensurate with its cost? The question is often asked, why is it that before Independence, when opportunities for scientific research were limited, the country produced some extraordinary scientists, yet when support for science has increased greatly, there seems to be a dearth of really outstanding scientists. This criticism may not be entirely just, but there is no getting away from the fact that there exists a certain feeling of discomfort with the state of affairs. Paradoxically, the loss in quality may in fact have resulted from too rapid an increase of support to science. When opportunities were restricted, educational and scientific institutions were compelled to be selective. With large numbers entering the profession, science faces the danger of becoming an employment rather than a calling.

Large scale governmentalization of science has inevitably led to its bureaucratization. Nehru had hoped that the association of scientists with the government will improve the working of government. In actual fact there has been a large scale reverse flow of rules, habits and mental attitudes from government departments to scientific institutions. The climate of many of our institutions is rule bound, hamstrung and age-ridden.

Lastly, have we succeeded in spreading scientific temper in society? I will not attempt to answer this last, obviously rhetorical question. Much as one would like to believe otherwise, looking at the socio-political and cultural scene around us, the impression is inescapable that Nehruvian ideas are somewhat on the decline.

The seeds of our failure

It is obviously unfair to blame Nehru for all our failures. Each society must carry on its science and its government within the constraints imposed by its history and culture. Society affects science nearly as much as science affects society. Nehru remarked somewhere, in relation to the destructive role of modern science, "We should not blame all our ills on science, we should improve ourselves." I could say the same for Nehru's science policy. We should not blame Nehru for all our failures. The harder question is, how does one improve "oneself"?

But there remains an interesting question. Were some of the failures of Nehru's policies inherent in his own ideas? Are these failures merely of administration and implementation or are some large issues involved?

For instance, was Nehru's faith in the power of science and technology as a universal panacea somewhat romantic and naive? Nehru's enthusiasm for science was, in later years, tempered by his realisation of its destructive potential and his concern for the spiritual side of man. But Nehru's beliefs had crystallised when economists and social thinkers did not quite appreciate the physical limits to growth and the destructive consequences of uncontrolled development.

Was Sir C.V. Raman right in his fear that the pursuit of big science will lead to a decline of the scientific spirit? Was loss of individuality an inevitable outcome of the large scale governmentalization of science that Nehru encouraged?

Why are there no men like Raman and Saha and Bhabha amongst us? Has science been grown too fast without paying attention to quality? Are we doing our science in wrong places and wrong ways so that we are driving the true men and women of science away?

Such question, are often raised during discussions on the state of our science. The subject is somewhat euphemistically described as "the health of science". The debate is more candid in public media. Nehru was not unaware of these problems and some of the dangers of his policies.

He asked, why is our science not as good as the best in Europe and America and noted that "we would rather sacrifice, it seems to me, rather cover up quality with quantity, and with the result that quality is sometimes lost in quantity, and yet science, obviously, depends much more on quality than quantity". (Science Congress, Delhi, 1963)

And of the abilities of his government he had the following to say:

> "You have to change your institutions so that they may function in a scientific way and not merely according to hierarchies and the like. That applies to the Government too — very difficult to change, because the apparatus of Government is built on seniority. People talk of the appeal of merit, but merit has no chance against seniority". (Ceylon Association for the Advancement of Science, Colombo, 1962)

"Governments may be good and may be bad but governments normally are very slow and the only thing that moves them is some immediate public outcry that affects their future directly. Therefore, I should discourage among the scientists a reliance always on what government may or may not do. (Science Congress, Delhi, 1947)"

He sympathised with the predicament of the scientist. "You may, of course, you do serve your various States in various capacities, and sometimes your abilities are taken advantage of by the State for purposes which probably you do not approve of or admire, yet it is difficult for you to do something about it". (Science Congress, Bangalore, 1951)

What about Nehru's idea of marrying politics to science and science to spirituality. He had the following to say:

"I am not sure that I would like the scientist to develop the politician's mind or approach to things. Now a politician's mind — I am not talking in any terms of criticism of the politician or the scientist, but they are different approaches. And I am not talking naturally of immature or indifferent scientists or the immature or indifferent politicians. I am talking about more or less competent politician and the more or less competent scientist. A politician seeks not so much the why of things as a scientist, but he wants to get somewhere. He wants to achieve certain results in the political field, in the economic field, in other fields. If a scientist's approach and a politician's approach were joined all together, then we get something". (Science Congress, Hyderabad, 1954)

But Nehru was aware that scientists are not saints and like politicians are tempted by power.

The power of Nehru's ideas

The relation of science to society has many aspects. Some are short range; these usually occupy the minds of politicians, administrators and technologists; others are long range; these evoke the interests of historians and philosophers. The remarkable thing about Nehru is that, although an active politician, his primary concerns were long term. He was too good a historian not to know that the forces of history are impersonal.

"None of you have the power to control the world's destiny, just as the politicians have not either. They may be leading persons in States or having the governance of large numbers of human beings in their hands and you may read about them in newspapers which give big headlines to the notorious politicians of the day. But the fact of the matter is that, however big they may appear in headlines, or in their offices, the forces that work in the world are infinitely greater than they and before them sometimes they almost appear as puppets unable to control not even themselves, leave out the forces." (Science Congress, Hyderabad, 1954)

Nehru was an optimist and I would like to conclude on a note of optimism. In humans, individuals as well as groups, a sense of well-being and optimism is essential for our minds to function well. If we have to survive, we must have faith in future. It seems to me, one needs optimism for the same reasons that one needs religion. If we take a long term view, and I should stress long term, Nehru's ideas give us reasons for optimism. The failure of his policies are for the short run. In the long run I believe that the seeds of the scientific-industrial revolution that Nehru has sown are the chief guarantee for India's survival.

I would like, at the end, to turn once again to the inimitable Jawaharlal.

"In the final analysis the biggest thing we have to do is to pull our people — a great majority of them — out of the tremendous rut of life which they have fallen into, the static condition, the unchanging condition in which they have lived — ploughing the same way they ploughed 2000 years, may be 3000 years ago, carrying on the same thing. That is the worst thing that had happened in India."

"We must realise that in the next few generations, may be a little more or a little less, vast changes will come, changing the way we live and the way we live affects the way we think.

Are we therefore at the dawn of a new civilization or is it the twilight of the old? Do we see around us in all this toil and trouble, the birth pangs of a new order, or something resembling the death agony of the old? I do not know. But given to a measure of optimism, I hope it is the dawn or the birth pangs and not the other. Anyhow, it is going to be something different. It is not going to be the same thing carried on." *(Discovery of India).*

Therein lies the force of Nehru's ideas.

Science and Peace

Y.S. Rajan

The subject is extremely complex and is not easily reducible to simple solutions such as 'developmental paradigms', 'more of international cooperation in pure and basic science', 'stoppage or reduction of defence oriented researches', etc. This brief article attempts to express some of the complex interconnected ideas in a 'telegraphic' form. The intention of the author is not to discourage some bold experiments or local actions aimed at 'peace' by painting a global scenario. But the thesis is that achievement of enduring 'peace' for the human society as a whole is not an easy task. We are yet to understand many aspects of the whole issue. The role scientists can play is indicated in various parts of this article.

What is science?

Science is a powerful tool used by the human being for 'cognition' of reality. There are many complex questions about the human understanding of nature. Science and scientific methods are based on the firm belief in the existence of objective reality and the human capability to organise it. Science will thus continue to be a venture of human curiosity and desire for knowledge. Besides this purely 'knowledge acquisition' facet of science, it is also pursued because such a pursuit can lead to the capability for 'control' — of production forces, environment, etc. In such a facet, science is considered as a directly productive force in development about which Prof. Sukhamoy Chakravorty spoke during the 77th Session of Indian Science Congress. He brought out the existence of two models: the 'Ionian model' (more of the pure knowledge acquisition-type) and the 'Bell-lab model' (more of the utilitarian-productive-force in development type).

One cannot easily conclude that the 'Ionian models' of pursuit of science will be inherently 'peaceful' and that the 'Bell-lab' can lead to peace. One reason is that they are not two separate immiscible channels of activities. Knowledge available to a society — whether it is internally or externally generated — and the 'utilitarian' activities devised from such knowledge interact in a very complex manner. Many other societal parameters such as economics, politics, culture, etc., come into play. A society which perceives, threat from some other society may use the new (and old) knowledge to remove the 'threats'. Such actions can even take aggressive forms.

What is peace?

We can now attempt to understand the concept of 'peace'. It cannot be interpreted merely as 'non-violence', 'abjuring of arms', etc. Any deep contradiction (real or perceived) between individuals, between groups of individuals (society) or between social groups and a nation-state or between nation-states can lead to disturbance of 'peace' immediately or later. Therefore, the search for 'peace' has to be done at all the levels:

- individual level
- societal levels (of group levels)
- national and trans-national levels

These three levels are not exclusively separate levels. All the three levels interact in a complex manner and there can also be strong and/or steady temporal variations within these levels and in the interaction between these levels.

For the search for peace at the individual level, the 'subjective realities' of the individual concerned (which is no doubt affected also by social-group-perceptions or the activities of the nation-states) will dominate over mere objective criteria (for example, human peace at this level cannot be standardised at literacy level, nutrition level, health index, etc). For the search for peace at the society or group level, different aspects of social psychology come into the picture; there are also historical aspects (real and perceived). Sometimes, in the histories of societies, individual psychologies or propensities of some 'opinion leaders' have played a crucial role. Somewhat similar is the situation at the nation-state level. Between the nation-states, many other political, geo-political, military-science aspect, etc., come into play.

Security perceptions of modern states

Assessment of the security-perceptions in the modern nation-states is no longer an easy judgment by the rulers (politicians, bureaucrats, technocrats, military professionals, etc). In this era of nuclear weapons, missiles, surveillance and communication satellites, etc., many abstract, mathematical concepts, such as the amount of overkill, range or reach of destruction, circular error probabilities, reaction times, etc. Method of dealing with such abstract models has a deep impact on impersonalising the decision-making process. Therefore, the modern nation-states have to be always prepared for the 'war games' played through such conceptual models and be ready with physical preparations to meet at least some of the possibilities predicted by the models. This is one of the reasons for heavy expenditure on military systems all over the world. It should not be forgotten that most of these (mathematical) models assume implicitly that 'peace' is nothing but a 'balance of forces'. That is the reason any reduction in arms is to be backed up by complex set of 'national technical means of verification'. Reductions in arms or aggressive manoeuvres may not be accepted, if there is a strong feeling in one of the nation-states that the agreed actions cannot be 'verified'.

Science and modern national security systems

The contributions of modern science not only to the aggressive weapons and delivery systems but also to the whole gamut of verification and command-control systems, etc., are enormous. One cannot easily dismiss these contributions or involvements as the products of the nexus between military and industrial complexes. Partly they are the outcome of the human history of conflicts, adventures, search for thrills, etc. Since many of the micro level aspects of the complex modern systems pose considerable scientific and technological challenges, the thrills obtained by the scientists working on these items cannot also be easily ignored. Purely at the individual and subjective levels, the search for the solutions to the scientific problems posed by these systems can be 'beautiful' (search for beauty). Lastly, we should not also ignore the fact that a large number of human beings in the modern societies depend on the 'non-peaceful work' not only from 'bread' but also for the high levels of affluence (satisfaction of hedonistic impulses).

Some perceptions of peace of the individual and social levels

Addressing now some aspects of perceptions of 'peace' at the 'individual' or the 'society' level one can delineate a number of serious differences. For some persons or groups any change (any disturbance of status quo) is a disturbance of 'peace'. For some groups or persons, the rules of such a change — the codification into the 'rule of law' — is more important. In another extreme, 'social delinquency' or 'terrorism' is considered as the 'fifth freedom' (Noam Chomsky). One has to understand this entire spectrum of perceptions and its complexity to find solutions. In some cases, one can resolve the 'conflicts' by resolving the 'contradictions' or by developing symbiotic relations.

Some solutions from science & scientific methods

The overall betterment of life of the human beings (albeit with many deficiencies in terms of its distribution) brought about by science gives us the hope that it is possible to reach the benefits to the whole of humanity. Of course, it is not assured automatically or by mere mechanistic extrapolations. This is where scientific methods especially involving social sciences will be of help. I would like to quote, in this context, the Presidential Address given by Prof. Prabha Gupta in the section on 'Psychology and Educational Science' during the 77th Session of Indian Science Congress. For example, it is pointed out that psychological approaches to 'peace' include: effective communication; improved self-esteem; providing good models; effective discussions, etc. Modern informatics systems are capable of assisting in this process by providing 'extended tools' to the social scientists.

Even while the findings of social sciences for achieving peace in the human society — especially at the individual and social levels — are very important. It is also necessary to keep in mind that some of the findings of modern biology (about genetic codes) may also have some relevance in dealing with individual and social levels. While suggesting this, I am not unaware of the fears expressed by Aldous Huxley in his writings such as 'Brave New World' or by Jeremy Rifkin in his recent book 'Algeny'.

But I don't think that we, as human-beings, will ever be able to run away from 'knowledge' and its 'derivatives' — whether these are the products of 'science' or not. 'Science' may not by itself (as we understand it now) be able to provide all solutions for the problems of humanity. 'Science' has to interact with the human society (as an important human creation), as an integral part of it. There will be new fears, unresolved questions, new solutions and new problems and so on. 'Scientific methods' have a role in this 'never ending' search for balance, beauty, equality, peace, adventure, etc.

In this context, it is worth noting that as people grow (or change) from one level of existence to another their perceptions of life, about their society and about themselves, change too. Therefore, possibilities of return to the 'romantic visions' of earlier societies are very remote. While such nostalgic feelings may have an impact on an individual or a society, it should be realised that by and large changes in human societies are irreversible and the new solutions have to be found keeping the totality of new interconnected world in mind. Scientific methods with their stress on observations, empirical studies and readiness to accept new theses and hypotheses, have a useful role to play in finding such new solutions.

Nexus between commerce and science — intellectual property

Let us address another important feature of modern science: the growing nexus between 'commerce' and 'science'. In a way it is a natural consequence of the 'Bell-lab' model of development. 'Market force', 'technology push' etc., have led to the insistence and mechanisms for the protection of Intellectual Property Rights (IPR). 'Scientific knowledge' is as much a property to be protected by law like 'Land', 'Capital', 'Equipment', and 'Machinery'. There are, therefore, new classes of 'possessors'. Let us not forget that many scientists belong to this 'new class' even while a large number of them are mere (well) paid workers in the 'production systems' of this 'new property'. With the globalization of economics and markets in the modern world, many trans-national companies (TNC) create environment for the production of this new property (of scientific and technological knowledge) in their well protected laboratories and factories, and protect the 'property' by various legal means and business strategies. New cost of new scientific inputs in this competitive game has grown tremendously in, recent years leading to developments of 'dual use' technologies (i.e, those which have application in 'defence' and 'civilian' sectors). Despite severe protection of intellectual properties, there are many mechanisms (including sale of products and services even for military systems) through which 'scientific knowledge' created in one place interacts with that created elsewhere. A better feature in one aircraft or an energy source even when described in 'functional terms' makes another firm or country think about it and achieve those specifications and if possible exceed them. Thus, even while 'science' is planned (often secretly in advance) in different places in accordance with the perceptions of the planners, there are a number of inter-connecting forces which operate on the 'scientific knowledge' generated in different parts of the world. Competition among the pure scientists for 'recognition' 'being first', etc., also leads to secrecy in planning and conduct of experiments. Modern world has become very complex, thanks to the modern science — there is no (easy) getting away from it. Totally insulated developments free from outside influences have also been proven to be impossible in today's world. Therefore, purely 'nationalist' approaches, howsoever emotionally appealing, may not withstand the global influences. This does not mean 'self-reliance' is impossible; one has to find new solutions and meaning for self-reliance.

Conclusion

In the search for peace, many of these complex aspects and interconnections have to be kept in mind. While scientists can and should help in the process of searching new solutions, one should not ignore the fact that scientists are also human beings and often would be in one or other side of a conflict, even without their knowledge. Modern scientific knowledge demands too much of a specialisation from an individual — so much so that he or she can very often be ignorant of its inter-connections with the rest of the 'system'. At the same time, concepts and the language of modern science are difficult and those untrained in 'science' would be 'lost' in understanding them or may derive simplistic or wrong conclusions. It is, therefore, necessary for many of the practitioners of modern 'science' to make many people of the society (and nations) understand the complex inter-connections and also learn from others to question some aspects of growth of 'science'. The First Pugwash Workshop on Non-military Mansions of Global Security held on November 24, 1989, addressed many inter-connected issues of heavy arms-race, environmental disasters and under-development. Development of larger loyalties, especially among the more fortunate sectors of society, is an important aspect. I would suggest that 'scientists' who know certain aspects of the modern human knowledge better should refrain from offering 'simplistic' or 'populist' solutions to the real and pressing problems. Modern science has brought the humanity to a stage which was described by Professor Yash Pal in his presidential address at the 77th session of Indian Science Congress as: "We are coming to a stage where there cannot be islands of development. The world is tied together in this box we call the earth....". 'Science' has the potential to help to unravel the inter-connections and offer solutions, if various societies aspire individually and collectively to achieve a peaceful world.

Recent Developments in Earthquake Prediction

Harsh K.Gupta

Earthquake is one of the worst natural calamities. As a conservative estimate, earthquakes have claimed more than 50 million human lives in the recorded history and inflicted damage worth several billions of dollars. As far as loss of human lives is concerned, the China earthquake of 1556 was the most damaging having claimed more than 8,00,000 lives. The second most damaging earthquake, also occurred in China recently on July 27, 1976 at Tangshan, claiming an estimated 5,00,000 lives. The third most catastrophic event occurred in 1737 in Calcutta killing an estimated 3,00,000 human beings. The memory of the Armenian earthquake of December 7, 1988 is fresh in our minds. According to a recent report (EOS, 1989) over 25,000 human bodies had been recovered and many more may have died.

Because of continuous increase in urbanization, with the passage of time, an increasingly larger human population is threatened by earthquakes. While in the year 1955, less than a 100 million people lived less than 200 km away from a possible magnitude 7 earthquake, this number has grown to 200 million now and will be about 400 million people by the year 2015.

Historically, earthquake prediction had practically no scientific status and was considered at par with alchemy. However, presently, we are undergoing the exciting phase of development of techniques, observational facilities and theoretical and practical investigations aimed to predict earthquakes. Careful observers have reported various physical phenomena precursory to earthquakes for many centuries. Seismologists in the past tended to discount most of this reporting. However, the measurable physical precursors are now being carefully investigated. The plate tectonics framework, developed in the sixties, provided physical basis for comprehending the spatial and temporal recurrence patterns of global earthquakes and created optimism and credibility for earthquake prediction in the scientific community.

What is an earthquake prediction?

U.S. National Academy of Sciences (1976) has given the following definition: "An earthquake prediction must specify the expected magnitude range, the geographical area within which it will occur, and the time interval within which it will happen, with sufficient accuracy so that the ultimate success or failure of the prediction can be easily judged. Only by careful recording of failures as well as successes can the eventual success of total efforts evaluated and total efforts be charted. Moreover, scientists should also assign a confidence level to each prediction....even low confidence predictions should be considered and evaluated."

Techniques of earthquake prediction

The first scientifically well documented prediction of earthquake was made at the Blue Mountain Lake, New York using spatial and temporal variation of ts/tp relation in 1973 (Aggarwal *et al*, 1973). The successful prediction of Haicheng earthquake of magnitude 7.3 on February 4, 1975 in China (Cha Chi-Yuan, 1976) and saving of an estimated 1,00,000 human lives has raised hopes that through a multi-disciplinary approach it is possible to go through various stages of earthquake prediction (long term — lasting several years, to medium term, short term and immediately prior to earthquake) leading to a successful prediction of impending earthquake. Although earthquake prediction has been taken up seriously only about a quarter of a century back, excellent reviews.and books have appeared dealing with the techniques of earthquake prediction (for example Rikitake, 1976; Asada, 1982; Simpson and Richards, 1981; Gu Gongxu and Ma Xinguan, 1984; Mogi, 1985; Iyer, 1988 and others). The scope of the present paper does not permit going into any details of these techniques. However, we will like to mention that Iyer (1988) has provided an excellent abbreviated version of various techniques currently in use. Table 1 lists the precursors which have been observed before earthquakes (Rikitake, 1982).

The structure of the crust of the earth being heterogeneous, the stress concentrates mechanically at singular points. As the external stress increases, local fractures and seismic or aseismic slip occurs, not along the fault plane alone, but also at widely distributed stress concentration points or positions of the crust causing micro-earthquakes, crustal deformation, water level changes, etc.; sometimes at places far removed from the focal region of the impending earthquake.

Table 1
Precursor observed before earthquakes
(Rikitake, 1982)

Precursor	No. of Data
Land Deformation	30
Tilt and Strain	89
Tidal Strain	2
Foreshock	83
Anomalous Seismicity	14
b-Value	12
Microseismicity	4
Source Mechanism	7
Hypocentral Migration	4
Fault Creep Anomaly	2
Vp/Vs	50
Vp and Vs	19
Geomagnetism	6
Earth Currents	17
Resistivity	32
Conductivity Anomaly	3
Radon	12
Underground Water	2
Oil Flow	3

Therefore, the spatial and temporal distribution of precursory phenomenon could be related to the degree of heterogeneity of the region (Mogi, 1984). A simple asperity model to explain precursory seismicity patterns has been proposed by Kanamori (1981). According to this model, a fault is constituted by a strong asperity and is surrounded by sub-faults. When the stress level is low, the sub-faults produce the background seismicity. With increase in tectonic stresses, more sub-faults fail causing the precursory swarms. Seismic quiescence occurs when most of the subfaults are broken, resulting in a highly stressed asperity decoupled from adjacent rock volumes. Finally, the asperity breaks causing the main shock.

Prediction of the Haicheng earthquake

Following is a summary of the four stages which the Sino scientists and public went through before successfully predicting the Haicheng earthquake of February 4, 1975 (Cha chi-Yuan, 1976).

1. First stage (long term 1970-73)

The objective was the commencement of work regarding regional studies of earthquakes, including historical earthquakes, field observations of major faults in Liaoning province, repeat measurements for crustal deformation, magnetic and D.S.S. profiling, setting of seismic stations, and observation of topographic changes at several places. These investigations revealed increase in seismicity. Sea coasts of Liaoning and Lioatung districts were suspected targets, since several 3 to 4 magnitude earthquakes occurred there. On the basis of these results it was concluded that Liaoning and Liaotung had greater probability of an imminent earthquake compared to other districts.

2. Second stage (medium term, later half of 1973 to October 1974)

The objective was to reinforce observations of precursory phenomena in the earthquake prone area bounded during the first stage through deploying portable means of observing crustal deformation and magnetic field changes, increasing the number of fixed observatories, training of the public in detecting and reporting short and medium term precursory phenomena, and obtaining support of all local Governments and party committees.

These investigations resulted in detecting a cumulative crustal movement, 20 times the normal annual value for September 1973 to June 1974 period south of Chinchow fault. An increase in the absolute value of the magnetic field was observed in Lutai district, indicating substantial crustal movement and increase in stress field. Remarkable rise in the sea level was revealed at Liaotung bay since 1973. The observatories recorded a 5-fold increase in micro-earthquakes.

The National Earthquake Bureau analyzed all the data and concluded that an earthquake of about 6 magnitude would occur in 1-2 years time. Consequent to this conclusion, earthquake management sections were set up and the value of public participation in earthquake prediction was publicized. By November 1974, prediction forecast points (such as underground electric currents, inclinations, ground-water and animal behaviour observation points) surrounded the seismic observatories. There were 232 such points in Yingkao city alone. Earthquake groups in each city exchanged data and held periodic meetings. Increased crustal movements, seismicity geomagnetic variations etc., and its concentration in Liaoning district was noticed and the area for a possible earthquake was further restricted.

3. *Third stage (short term, November 1974)*

Efforts at observing short term precursory phenomena were intensified. It was observed that the snakes hibernating came out of their burrows and were frozen stiff. Water levels in wells were found to change. The radon content increased by 20 to 40 percent. On December 22, a 4.8 magnitude earthquake occurred 70 km north of Haicheng. In middle January 1975, the National Earthquake Bureau again called a conference, examined all the data and predicted an earthquake of magnitude 6 in the first half of 1975.

4. Fourth stage (immediately prior to earthquake, January 1975)

The objective was to reinforce the various observations and catch the "Big One". There was an emergency meeting on January 28, and the officials and public were alerted and instructed on earthquake disaster prevention, particularly in mines, factories, etc., and various duties concerning prevention and minimization of damage and loss of lives were assigned. In the beginning of February 1975, the anomalous behaviour of animals increased. Thc anomalous phenomenon observed in the well water advanced from SE and NW directions towards Yinkao-Haicheng. Beginning on February 2, an abrupt change in electrical current was observed. A level-meter placed in a cave showed an accelerated break in the SW direction. The hourly number of shocks rose to 60, located 20 km away from the Yingkao earthquake station. Yingkao earthquake observatory analyzed all the data and predicted that an earthquake would occur in the next few days. Two earthquakes of magnitude 4.7 and 4.2 occurred on the morning of February 4. At 14:00 hours on February 4, instructions were issued through emergency telephones to get ready for the earthquake and all the earthquake disaster prevention troops conducted preventive measures in the Haicheng area. At 19.36, on February

4, the 7.3 magnitude Haichang earthquake occurred. Successful prediction and timely evaluation saved an estimated 1,00,000 human lives.

Parkfield Earthquake Prediction

As summarized by Bakum (1987), characteristic magnitude 6 earthquakes had occurred on the Parkfield section of the San Andreas Fault, California in 1881, 1901, 1922, 1934 and 1966. Very detailed studies by U.S. earth scientists have led to the conclusion that the next characteristic earthquake shall occur before 1993. The anticipated magnitude is 6, which might trigger, or grow into, a magnitude 7 shock on the adjacent northern Carrizo Plains section of the San Andreas Fault (Sieh and Jahns, 1984). As a prediction experiment, the main objective of Parkfield experiment is a detailed description of the final stages of the earthquake preparation process, which should help evaluation of feasibility of intermediate and short term earthquake prediction elsewhere. A secondary goal of the experiment is to issue a short term warning (minutes to days) to the agencies responsible for public dissemination of any warning and activation of the relevant disaster mitigation efforts. The Parkfield earthquake experiment has provided a unique opportunity to capture an anticipated earthquake using an extensive network of monitoring devices.

The Loma Prieta earthquake of October 17,1980

The 7.1 magnitude Loma Prieta earthquake of October 17, 1989 in northern California caused 61 confirmed fatalities, 3,100 injuries, rendered more than 10,000 homeless, and the property losses are estimated to be about 6 billion dollars (Staff of U.S.G.S. 1990). This is considered to be the most expensive natural disaster in the USA. According to the scientists of U.S.G.S., this was an anticipated event. All along the San Andreas Fault, sites of future potential earthquakes have been identified, and probability estimates of their occurrence over the next one or two decades have been made. According to this compilation, the site of Loma Prieta earthquake had a 30 percent probability of occurrence over the next two decades. It may be noted that the site of the predicted Parkfield earthquake has a 90 percent probability of occurrence before the end of 1991.

Tokai earthquake prediction

In Japan, an earthquake prediction programme was started as a national project in 1965. A programme to constantly monitor and assess various parameters was formally initiated in 1978 for forecasting the major earthquake M~8 that is expected to occur in the near future in the Tokai district of Central Houshu Island (Mogi, 1986). The two main reasons for the Tokai earthquake to occur are: 1) large earthquakes have often occurred in this region in the past, but no such event has occurred for many years, and 2) crustal strain has steadily increased in this region for past several years. The main issue as to when this earthquake will occur remains unresolved. However, Japanese seismologists have deployed extensive facilities for observing precursory phenomena, and it is hoped that it would be feasible to issue a short term prediction of the expected Tokai earthquake. The Japan Meteorological Agency responsible for short term prediction of the expected Tokai Earthquake has installed a system called Earthquake Phenomena Observation System (EPOS), that can process 512 channels of seismic data and 512 channels of other data (for example strain, sea level, tilt etc.) on an on-line real-time basis (EOS, 1988). All EPOS data, including real time data can be displayed on a large screen for the earthquake prediction committee to assess the situation and issue necessary instructions on the basis of most recent observations.

Earthquake prediction studies in India

The Himalayan frontal arc, flanked by the Chaman fault in the west and the Arakan Yoma fold belt in the east, constitutes one of the seismically most active intra-continental region in the World. Four great earthquakes ($M \geq 8$) occurred during the past nine decades (1897, 1906, 1934 and 1950) claiming a large number of human lives and causing much destruction. Figure 1, updated from Chandra (1978) depicts epicenters of earthquakes of $M \geq 7$, and the ones claiming human lives (for some of them, being historical, instrumental magnitude is not available). Another region, of immense interest is the world famous site of reservoir induced seismicity (RIS) at Koyna, Maharashtra where the largest RIS event occurred on December 10, 1967, having a magnitude of 6.3.

Some of the earliest earthquake prediction related studies were carried out at the Regional Research Laboratory, Jorhat, a CSIR Laboratory, when Khazanchi and Dutta (1978) statistically estimated the return periods of large magnitude earthquakes in the north-east India region. A symposium organized by the India Meteorological Department in 1978 reviewed the earthquake prediction related work (*Mausam*, 1979). Another recent symposium reviewed the current research efforts (Guha and Patwardhan, 1988). The techniques used include spatio-temporal variation of

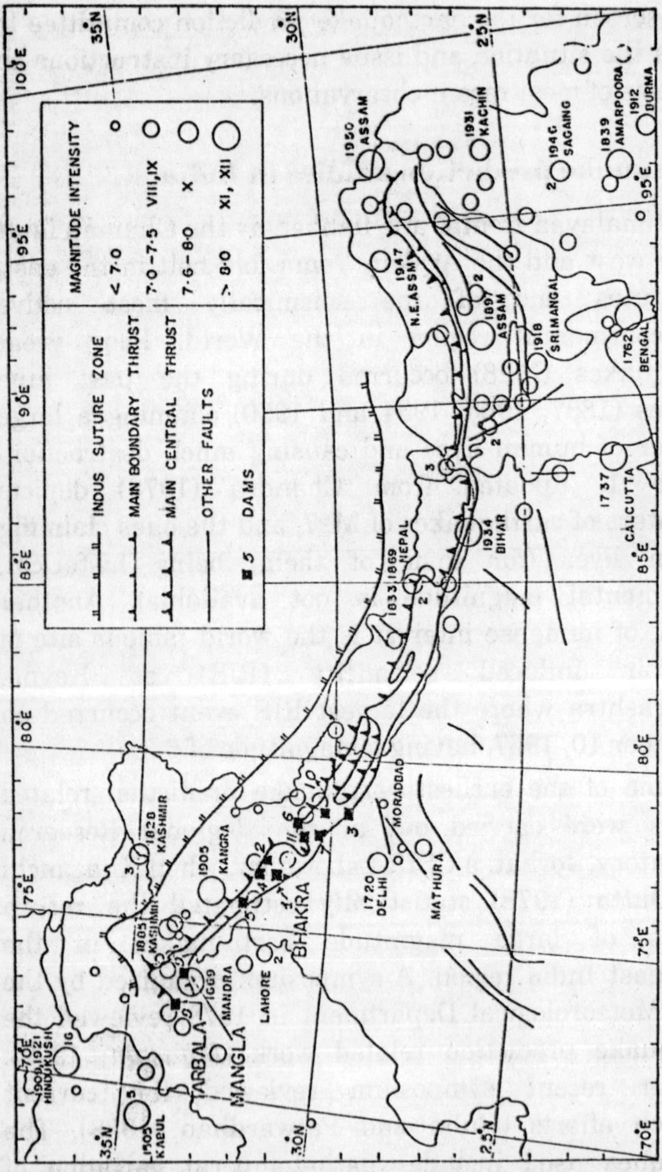

Fig.1 An epicentre map of Himalaya and vicinity showing earthquakes of M 7.0 as well as earthquakes that claimed human lives (updated from Chandra, 1978).

seismicity (Khattri and Wyss, 1978; Srivastava and Chaudhury, 1978; Srivastava 1979, 1983, 1988; Gupta and Singh, 1986; Singh *et al*, 1982 and others); micro-earthquakes (Khattri *et al*, 1983; Khattri and Tyagi, 1983; Gupta *et al*, 1984; Saha, 1988; Rastogi *et al*, 1986, b and c and others); statistical methods (Srivastava and Dattatrayam, 1986; Sarmah, 1988); 'b' values (Srivastava *et al* 1987; Gupta and Pagare, 1988; Guha, 1980 and others); levelling and geodetic measurements (Arur and Singh, 1988); radon emission (Rastogi, 1988; Ghosh *et al*, 1988; Virk, 1988, Gupta, 1988); teleseismic P-delays (Gupta and Singh 1980; Srivastava and Chaudhury, 1979); geomagnetic and geoelectric variation (Singh, 1988; Arora, 1988); resistivity (Saha, 1988) and animal behaviour (Deshpande 1988). This list is by no means exhaustive, and the scope of the paper does not permit going into the details and limitations of these investigations. The point we wish to make is that some initial work has been carried out to investigate a number of known earthquake precursors in India.

Himalyan frontal arc

Srivastava (1988) has summarized the operation of special seismic networks operated in Himachal Pradesh and the northeast India, as well as precursory phenomena observed in these regions. However, for long term prediction of great earthquakes, and to identify the zones of future earthquakes, it is useful to develop a suitable model for a region. The model, that we prefer, for the Himalayan frontal arc is the one developed through the compilation of Seeber and Ambruster (1981), Khattri and Tyagi (1983) and others. Structurally, Continental subduction along arcs, and hence the concept of seismic gaps holds good for it. Great earthquakes rupture the detachment in both cases. For the Himalayan frontal arc, the total length is about 1700 km, out of which 1400 km has been broken during the last four great earthquakes of 1897, 1905, 1934 and 1950, leaving a portion of about 300 km to be broken in a future great earthquake (Figure 2). During the century before 1897, there were possibly three great earthquakes, i.e., 1885 in Kashmir, 1833 in Western Bihar and 1803 in Uttar Pradesh. The extent of rupturing for these earthquakes cannot be constrained. However, according to Seeber and Ambruster (1981), the most likely unruptured gaps are those between 1803 and 1833 ruptures in Uttar Pradesh and between 1885 and 1905 ruptures in Kashmir. They further add that the entire Himalayan detachement would rupture in 180 to 240 years and repeat time of a typical 300 km rupture for a M 8 earthquake would be 200-270 years. It may be mentioned that there is a considerable difference of opinion : for example the 300 km long rupture inferred by Seeber and Ambruster for the 1934 earthquake is much longer than the estimates of Singh and Gupta (1980) obtained from the spectral ratio of surface waves travelling in opposite directions from the rupture. However, their model provides a good working model for further work.

Khattri and Wyss (1978) have demonstrated that major earthquakes in the north-east India region are preceded by quiescence. Between the meizoseismal areas of 1897 and 1950 great earthquakes, they have identified a seismic gap which could be the site of the next major earthquake in the region (Figure 3). However, Seeber and Ambruster (1981) are of the opinion that the 100 km long gap between the ruptures of the 1897 and 1950

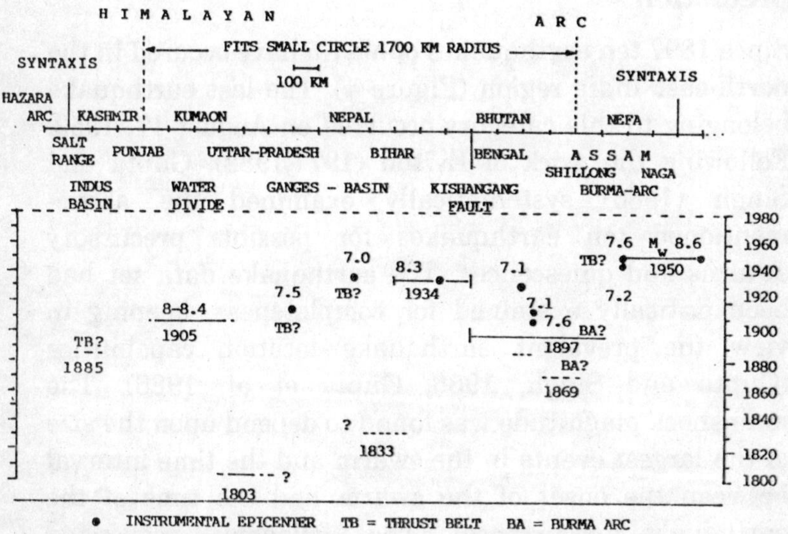

Fig.2 Space-time diagram of detachment rupture in the Himalayas are since 1800 (after Seeber and Ambruster 1981). The data before 1897 may be incomplete. The most likely gaps unruptured since 1800 are in Kashmir and Uttar Pradesh, between the 1833 and 1803 ruptures. The gap between the 1897 and 1950 ruptures was probably filled by the 1943 and 1947 earthquakes. If the 1833 and 1803 ruptures about, the entire Himalayan detachment, excluding Hazara and Kashmir where seismic slip may play an important role, has probably ruptured since 1800.

earthquakes may have been filled by the two large earthquakes of 1943 and 1947, and, therefore, this gap is a possible, but not a likely, site of the next great detachment earthquake.

Medium term Arakan Yoma Earthquake prediction

Since 1897 ten earthquakes of $M \geq 7.5$ have occured in the north-east India region (Figure 4). The last earthquake belonging to this category occurred on August 17, 1952. Following the work of Evison (1977,1982), Gupta and Singh (1986) systematically examined the above-mentioned ten earthquakes for possible precursory swarms and quiescences. The earthquake data set had been critically examined for completeness, keeping in view the prevalent earthquake location capabilities (Gupta and Singh, 1986; Gupta et al, 1986). The mainshock magnitude was found to depend upon the size of the largest events in the swarm and the time interval between the onset of the swarm and the time of the earthquake occurrence. The following regression equations relating the mainshock magnitude (Mm), the average magnitude of the largest two events in the swarm (Mp) and the time interval in days (Tp) between the beginning of the swarm and the mainshock were developed :

$$Mm = 1.37\ Mp - 1.41$$

and

$$Mm = 3 \log 10\ \ Tp - 3.27$$

Gupta and Singh (1986) pointed out that it is important to recognize the swarms and quiescence before the occurrence of the mainshock, and reported that they have identified one such area in the vicinity of

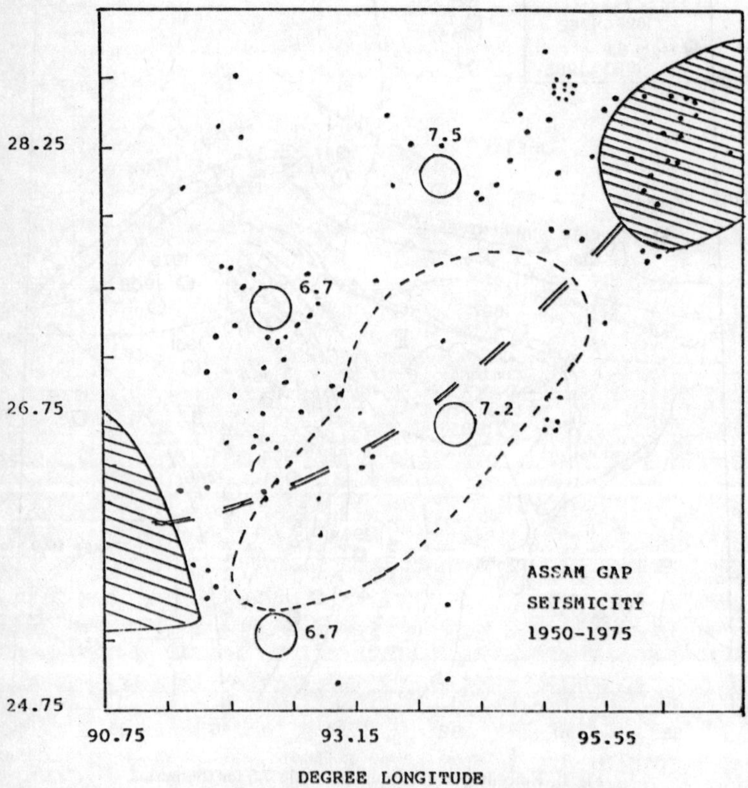

Fig.3 Bounded by the meizoseismal areas of 1897 and 1950 great earthquakes of northeast India, both of M 8.7, the Assam gap is the area within dotted lines identified by Khattri and Wyas (1978).

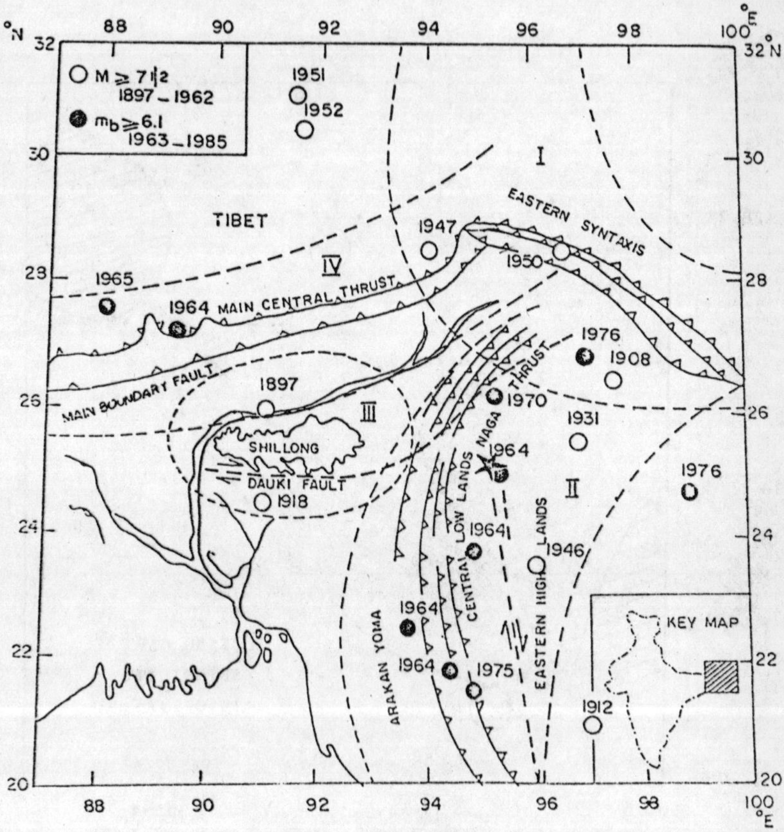

Fig.4. Epicentre of earthquakes of M≥ 7.5 for the period 1897--1962 (open circles) and $m_b \geq 6.1$ for the period 1963-1985 (filled circles) in the northeast India region (adapted from Gupta and Singh, 1986) Star: epicentre of M 7.3 August 6, 1988 earthquake.

India-Burma border region, where four earthquakes of mb6.1 occurred in a quick succession in 1964. To examine this swarm, an area bounded by 20° N and 26° N latitudes, and 92° E and 98°E longitudes was tentatively selected. Earthquake data from 1950 through 1985 were considered. Figure 5 depicts earthquakes of mb ≥ 5.0 and mb ≥ 5.5 where epochs of backgrounds, swarm and quiescence seismic activity are quite well identified. For example, the annual frequency of mb ≥ 5.5 events within this elliptical zone is found to be 0.4, 4.4 and 0.3 during the epochs of normal/background, swarm and quiescence seismic activity. A several fold increase in seismic activity during swarm, as compared to background and quiescence period, is obvious. On the basis of the spatial distribution of swarm events an elliptical area of 130,000 sq km was tentatively delimited as the preparation zone for the future earthquake. Gupta and Singh (1986) noted that since the beginning of the swarm, some 23 years had passed. The mp value for swarm was 6.6. Using these parameters, a forecast for a major events, as per parameters given in Table 2 was made. The focal depth of 100±40 km was estimated in view of the focal depth of the swarm events. The Tp for M ≥ 7.5 events varied from 11 years to 27 years. The two great earthquakes of 1897 and 1950 were preceded by swarms which began 23 years and 27 years before the main earthquakes respectively. Therefore, it was stated that if this forecast does not come true till 1990 (27 years after the initiation of the swarm) it should be considered as a false alarm. Gupta and Singh (1986,1987) have pointed out several shortcomings of their analysis. For example, they have not formulated rigorous criteria for identification of background seismicity, swarm and quiescence (such as those by

Michael and Toksoz, 1982). Another problem was the forecast being singular. Ideally several similar situations elsewhere in the world should be identified to examine the applicability of swarm hypothesis in forecasting earthquakes. However, the remarkable success of Gupta and Singh's (1986) medium term earthquake forecast based on observing precursory swarm and quiescence should provide impetus to similar investigation elsewhere.

Table 2

Earthquake parameters	Prediction (Gupta and Singh, 1986)	Occurrence NEIS (Preliminary determination)
Epicentre	21° N-25$^1/_2$°N	25/149
Magnitude	8 ± $^1/_2$	7.3
Depth	100 ± 40 km	90.5
Time	Feb.86-Dec.90	Aug. 6, 1988 (00:36:24.65 G.C.T.)

Earthquakes at Koyna

The continued seismic activity in the vicinity of Koyna dam in Maharashtra is a classical example of reservoir induced seismicity (Guha et al, 1974 :Gupta and Rastogi, 1976). The largest known RIS earthquake at Koyna occurred on December 10, 1967. This 6.3 magnitude earthquake claimed over 200 human lives, injured over 1500 persons and rendered thousands homeless. Earthquakes exceeding magnitude 5 occurred in November 1973 and September 1980. Significant work on RIS, relevant to earthquake prediction has been carried out at the National Geophysical Research Institute, Hyderabad. In a couple of pioneering studies

(Gupta et al 1972 a and b) criterion were generated to discriminate RIS from natural earthquakes of concerned regions. These are now applied internationally. Gupta (1983) has shown that earthquakes of magnitude of 5 and more occurred only when rate of reservoir filling exceeded 40 ft/week. However, this is a necessary, but not sufficient condition. In another study, Gupta and Iyer (1984) demonstrated that there appears to be a high probability of occurrence of a magnitude 5 earthquake when two earthquakes (foreshocks) of magnitude 4.0 occur within a fortnight. Changes in several precursory phenomena, such as 'b' values, seismic wave velocity anomaly, tilt etc. are reported by Padale et al (1979) before significant earthquakes in the Koyna region.

Concluding remarks

Earthquake prediction related work in India has just started. Most of the work, so far, is hind sight, where search is being made for precursors after the earthquake has taken place. However, this is an essential exercise to learn as to what to look for in different regions. Traditionally, in India, the India Meteorological Department and the Geological Survey of India investigated the earthquake phenomena. Over the last 25 years, several other institutions/laboratories have started earthquake related studies. The efforts of the National Geophysical Research Institute, Hyderabad and Regional Research Laboratory, Jorhat need a special mention. Between these two CSIR laboratories a multi-element seismic array has been put into operation in the north-east India region and several earthquake prediction related studies are being undertaken. Plans are also underway for satellite communication of earthquake data to make near real time analysis feasible, and for very long baseline radio interferometry measurements (VLBI) for inferring relative plate movements. (National Geophysical Research Institute, Annual Reports, 1986-87, 1987-88).

It appears that certain cycles of low and high seismicity characterise the Alpide belt. During the active phase from 1897 through 1916, fifteen earthquakes of M ≥ 7.7 occurred in the immediate vicinity of the Alpide belt. This was followed by a lean phase during 1917 through 1933 when no such earthquake occurred. Again there was an extremely active phase during 1934 through 1951 when 14 earthquakes of M ≥ 7.7 occurred in the immediate vicinity of Alpide belt. A quiet phase began in 1952, and till now only four such events have occurred. The occurrence of a few earthquakes in a short span of time during 1988, for example the India-Burma border earthquake of M ≥ 7.3 on August 6, 1988; the Hindukush earthquake of M ≥ 6.7 on August 6, 1988; the Bihar Nepal earthquake of M ≥ 6.7 on August 21, 1988 and the recent Armenian earthquake of M ≥ 7.0 on December 8, 1988, may mark the beginning of another phase of high seismic activity along the Alpide belt. It may be noted that after independence, several major civil structures have been undertaken in the vicinity of the Alpide belt in India. Non-occurrence of a major earthquake over the last 38 years in India has generated a false sense of security. The recent Bihar-Nepal earthquake was only of magnitude 6.7, and it caused widespread damage and claimed over 1000 human lives. It should be remembered that in the same region, the Bihar-Nepal earthquake of January 12, 1934 of M 8.4 claimed an estimated 11,000 lives. With the increase of population and civil structures Indian cities in the vicinity of Alpide belt and elsewhere are becoming more vulnerable to the hazards of earthquake damage.

The assessment made by Suzuki (1982) that no omnipotent method for successful prediction of earthquakes is known at present and further research in various approaches must be carried out before prediction is practically available, is still valid. Results of multi-disciplinary experiments to make short term prediction of earthquakes at Parkfield in California and Tokai in Japan will shape the future of earthquake prediction research.

Acknowledgements

This is an update version of an article "Earth-quake Prediction: International and Indian Scenarios" that appeared in "40 Years of Research - A CSIR Overview" in September 1989. Dr. A.P.Mitra, then Director General, CSIR kindly permitted reproduction of this article. Several colleagues urgently sent me their reprints/preprints for preparation of this paper. Mr. K.P. Sasidharan assisted in typing.

References

Y.P. Aggarwal, L.R.Sykes, J.Ambruster, and M.L.Sbar, Premonitory Changes in Seismic Velocity and Prediction of Earthquakes, *Nature,* **241**, 1973, pp.101-104.

B.R. Arora, *Tectonomagnetic Studies in India,: Earthquake Prediction-Present Status*, University of Poona, 1988, pp.53-62.

M.G. Arur and Singh, A.N., *Earthquake Prediction for Levelling Data : An Integrated Approach, Ibid,* pp. 67-73.

T. Asada, *Earthquakes Prediction Techniques*, University of Tokyo Press, 1982, pp.317.

W.H. Bakun, *Future Earthquakes, In: Reviews of Geophysics*, U.S. National Report to International Union of Geodesy and Geophysics, 1983-86, 25(6):1987, pp. 1135-1138.

R. Bilham, *Urban Growth, Earthquake Hazard and Space Geodesy*, Preprint, 1989.

U. Chandra, Seismicity, Earthquake Mechanisms and Tectonics along the Himalayan Mountain Range and Vicinity, *Phys. Earth Planet. Inter.*, 6, 1978, pp. 109-131.

C. Chi-Yuan, *General Conditions of Earthquake Studies and Actions in China, in Proceedings of the Lectures by the Seismological Delegation of the People's Republic of China*, prepared by National Aeronautics and Space Administration, 1976, pp. 5-11.

B.G. Deshpande, Earthquakes, Animals and Man, *Earthquake Prediction-Present Status*, University of Poona, 1988, pp. 63-66.

EOS: *U.S.-Japan Quake Prediction Research,* Report by Carl Kisslinger, Takeshi Mikuo and Hiroo Kanamori, 69, No. 52, 1988, pp.1672-75.

F.F. Evison, Fluctuations of Seismicity Before Major Earthquakes, *Nature* (London), 225, 1977, pp.710-712.

F.F. Evison, Generalised Precursory Swarm Hypothesis, *J. Phys. Earth*, 30, 1982, pp.155-170.

J.K. Ghosh, T.Krishnan, and T.S.Kutty, A Stochastic Model for Earthquakes and Prediction via b-Values, *Earthquake Prediction-Present Status,* University of Poona, 1988, pp.103-110.

Gu Gongxu and Ma Xinguan (Eds.), *On Continental Seismicity, and Earthquake Prediction*, Seismological Press, Beijing, China, 1984, pp.867.

S.K. Guha, Studies on Earthquake Precursors, *Earthquake Prediction-Present Status*, Poona University, 1988, pp.31-33.

S.K. Guha, P.D.Gosavi, K.Nand, J.G. Padale, and S.C. Marwadi, *Koyna Earthquakes (October 1963 to December 1973)*, Central Water and Power Research Station, Khadakwasla, Poona, India.

S.K. Guha, and A.M.Patwardhan, *Proceedings of the Symposium on Earthquake Prediction: Present Status*, University of Poona, 1986.

H.K. Gupta, Seismicity of North-east India Region and Earthquake Forecasts, *Ind.J.Geol.*, 59, 1987, pp.165-178.

H.K. Gupta, Induced Seismicity Hazard Mitigation through Water Level Manipulation : A Suggestion, *Bull.Seismol. Soc. Am.*, 73, 1983, pp.679-682.

H.K. Gupta, and H.N.Singh, Seismicity of north-east India region, Part:II, Earthquake swarms precursory to moderate magnitude to great earthquakes, *J.Geol.Soc. of India;* 28, 1986, pp.367-406.

H.K. Gupta, and V.P.Singh, Teleseismic P-wave Residual Investigations at Shillong, India, *Techonophysics*, 66, 1980, pp.T19-T27. 1980.

H.K. Gupta, K. Rajendran, and H.N.Singh, Seismicity of north-east India Region, Part I: The Data Base, *J. Geol. Soc. of India*, 28, 1986, pp.345-365.

H.K. Gupta, B.K.Rastogi, and H.Narain, Common Features of the Reservoir Associated Seismic Activities, Bull. *Seism. Soc .Am.*, 62, 1972a, pp.481-492.

H.K. Gupta, B.K.Rastogi, and H.Narain, Some Discriminatory Characteristics of Earthquakes near the Kariba, Kremasta and Koyna artificial lakes, *Bull. Seism.Soc.Am.*, 62, 1972b, pp.493-507.

H.K. Gupta, S.C.Singh, T.K.Dutta, and M.M.Saikia, *Recent Investigations of North-East India Seismicity, in Proceedings: International Symposium on Continental Seismicity and Earthquake Prediction*, Gu

Gongxu and Ma Xing-Yuan (Eds.), Seismological Press, Beijing, 1984, pp.63-71.

H.K. Gupta, and B.K.Rastogi, *Dams and Earthquakes*, Amsterdam : Elsevier Scientific Publishing Co., 1976.

H.K. Gupta, and H.M.Iyer, Are Reservoir Induced EarthQuakes of Magnitude 5.0 at Koyna, India, Preceded by Pairs of Earthquakes of Magnetude 4.0?, *Bull. Seismol.Soc.Am.*, 74(3), 1984, pp.863-873.

I.D. Gupta, and V.S.Pagare, b-Value As A Premonitory Parameter for Earthquake Prediction : An Example Study for Bhatsa Dam Region (India), *Earthquake Prediction-Present Status*, University of Poona, 1988, pp.75-77.

S.K. Gupta, Earthquake Prediction using Hydrogeo-Chemical Precursors, *Ibid*, 1988, pp. 135-139.

K. Hamada, *Prediction-oriented seismology, in Current Research in Earthquake Prediction*, Tsuneji Rikitake (Ed.), *Development in Earth and Planetary Sciences*, Part 02, Centre for Academic Publications, Japan, 1981, pp.221-300.

H.M. Iyer, Earthquake Prediction-How Close is it to Reality?, *Earthquake Prediction--Present Status,* Poona University, 1986, pp.1-11.

H. Kanamori, *The Nature of Seismicity Patterns before Large Earthquakes in Earthquake Prediction--An International Review*, Maurice Ewing Series 4: D.W. Simpson and P.G. Richards (Eds.), American Geophysical Union, Washington, D.C., 1981, pp.1-19.

K. Khattri, and A.K.Tyagi, Seismicity Patterns in the Himalayan Plate Boundary and Identification of the Areas of High Seismic Potential, *Tectonophysics*, 96, 1983, pp.281-297.

K. Khattri, and M.Wyss, Presursory Variation of Seismicity Rate in the Assam Area, India, *Geology*, 6, 1978, pp.685-688.

A.C. Khazanchi, and T.K.Dutta, *Seismic loads and Structural Safety with Particular Reference to the North-Eastern Region of India*, 3rd Symp. on Earthquake Engineering, Univ. of Roorkee, 1, 1978, pp.523-528.

Proceedings of the Symposium on Earthquake Prediction, March 8-10, 1978, *Mausam* New Delhi, 30, 1979, pp.173-403.

A.J. Michael, and M.N.Toksoz, Earthquake Swarms as a Long-range Precursor to Large Earthquakes in Turkey, *Geophys. J.R.Astr.Soc.* 68, 1982, pp.459-476.

K. Mogi, *Earthquake Prediction*, New York : Academic Press, 1985.

K. Mogi, Fundamental Studies on Earthquake Prediction, In: *International Symposium on Continental Seismicity and Earthquake Prediction*, Seismological Press, Beijing, 1984, pp.619-652.

K. Mogi, Recent-Earthquake Prediction in Japan, *Science* 1986, pp.324-330.

J.G. Padale, P.D.Gosavi, S.C.Marwadi and S.K.Guho, Feasibility of Earthquake Prediction in Koyna Region (India), *Mausam*, 30 (283), 1979, pp.281-288.

B.K. Rastogi, Patterns of Earthquake Precursor Parameters for Some Earthquakes. *Earthquake Prediction--Present Status*, University of Poona, 1988, pp.79-85.

B.K. Rastogi, C.V.R.K., Rao, R.K.Chadha, and H.K.Gupta, Microearthquakes near Osmanasagar Reservoir, Hyderabad, India, *Phys. Earth. Planet. Inter.*, 44, 1986a, pp.134-141.

B.K. Rastogi, B.R.Rao, and C.V.R.K. Rao, Microearthquake Investigations near Sriramsagar Reservoir, A.P., India, *Phys. Earth. Planet. Inter.*, 44, 1986a, pp.149-159.

B.K. Rastogi, R.K.Chandha, and I.P.Raju, Seismicity near Bharsa Reservoir, Maharashtra, India, *Phys.Earth.Planet.Inter.*, 44, 1986c, pp.179-199.

T. Rikitate, *Earthquake Prediction*, Amsterdam : Elsevier Scientific Publishing Co.,1976.

T. Rikitate, *Earthquake Forecasting Warning*, Dordrecht : D.Reidel Publishing Co., 1982.

S.N. Saha, Micro-earthquakes and Earthquake Prediction Oriented Studies in North-Eastern India; present status, *Earthquake Prediction-Present Status*, University of Poona, 1988, pp.247-254.

S.K. Sarmah, *Long Term Earthquake Prediction in Northeast India*, Ibid, 1988, pp.111-115.

L. Seeber, and J.G.Armbruster, Great Detachment Earthquakes along the Himalayan Arc, and Long Term Forecasting, *Earthquake Prediction--An International Review*, Maurice Ewing Series 4, American Geophysical Union, 1981, pp.259-277.

K.E. Sieh, and R.H.Jahns, Holocene Activity of the San Andreas Fault at Wallace Creek, California, *Geol. Soc. Am. Bull.*, 95, 1984, pp.883-896.

D.W. Simpson, and P.G.Richards, (Eds.), *Earthquake Prediction--An International Review*, Maurice Ewing Series 4, American Geophysical Union, Washington, D.C., 1981.

B.P. Singh, Geomagnetic and Geoelectric Changes Associated with Earthquakes, *Earthquake Prediction--Present Status*, University of Poona, 1988, pp.39-52.

D.D. Singh, and H.K.Gupta, Source Dynamics of Two Great Earthquakes of the India Subcontinent: The Bihar-Nepal Earthquake of January 15, 1934 and the Quetta earthquake of May 30, 1935, *Bull. Seismol.Soc.Am.*, 70, 1980, pp.757-773.

V.P. Singh, H.N.Singh, and J.Singh, On the Possibilities of Premonitory Searms for Three Sequences of Earth-Quakes of Burma-Szechwan Region, *Tectonophysics*, 85, 1982, pp.T21-T29.

H.N. Srivastava, Precursory Seismic Observation in Himachal Pradesh and Shillong Plateau, *Mausam*, 30, 1979, pp.289-296.

H.N. Srivastava, Forecasting Earthquakes, *Delhi: National Book Trust, 1983.*

H.N. Srivastava, Search for Earthquake Precursors in Himachal Pradesh and Northeast India, *Earthquake Prediction--Present Status*, University of Poona, 1986, pp.255-265.

H.N. Srivastava, and H.M.Chaudhury, Possibility of a Large Earthquake in 'Seismic Gap' of Himachal Pradesh, India, *Proc. Vth World Conf. Earthquakes Engg.*, Tokyo, 1978, pp. 353-359.

H.N. Srivastava, and H.M.Chaudhury, Regional Plate Tectonics from Himalayan Earthquakes and their Prediction, *Mausam*, 30, 1979, pp.181-186.

H.N. Srivastava, and R.S.Dattatrayam, Study of Return Periods of Earthquakes in Some Selected Indian and Adjoining Regions, *Mausam*, 37, 1986, pp.333-340.

H.N. Srivastava, R.K.Dube, and Hans Raj, Space and Time Variation in the Seismicity Pattern Preceding Two Earthquakes in the Himachal Pradesh, India, *Tectonophysics*, 138, 1987, pp.69-87.

Z. Suzukim, Earthquake Prediction, *Ann.Rev.Earth Planet Sci.*, 10, 1982, pp.235-256.

U.S. Geological Survey Report, *"The Loma Prieta, California Earthquake: An Anticipated Event"*, California, 1989.

U.S. National Academy of Sciences: *Predicting Earthquakes--A Scientific and Technical Evaluation with Implications for Society*, Washington, D.C.: National Academy of Sciences, 1976.

H.S. Virk, *Radon Monitoring and Earthquake Prediction*, Earthquake Prediction: Present Status, University of Poona, pp.157-162.

Monoclonal Antibodies and Abzymes off the Bottle

D. Balasubramanian

One of the exciting developments in biology in the last few years has been the fusion of the idiom of the enzymatic catalysis with that of immunity and antibody response. There are a few tens of thousands of enzymes present in nature, each of which catalyses a specific reaction of biochemical relevance. Yet, there are no enzymes readily available for many reactions of technical or synthetic chemical reactions, and it would be valuable to be able to have enzymes tailor-made or designed for a given reaction. On the other hand, antibody molecules come in millions of different types, each one interacting specifically with a given antigen molecule — but antibodies do not catalyse chemical reactions. It would be valuable fusion of the features if antibodies can be made to catalyze chemical reactions!

In both cases, what is of paramount importance is the intermolecular recognition that characterises the process. Recognition is so specific that distinction is made between two proteins that differ in but one amino acid residue, or between two molecules that differ in but one chemical bond, namely *stereospecific recognition*. Enzyme catalysis involves the grabbing of the *substrate* molecule by the enzyme, using a multiplicity of weak interactions. The stereospecifically bound substrate is now reacted upon by chemical functional groups in the enzymes and the reaction is completed rapidly and efficiently. The reaction products do not fit as well as the substrate did on the enzyme's binding site and hence are released. The enzyme is now ready for the next round of the catalytic acceleration of the reaction of a second substrate molecule. Stereospecificity of the substrate binding, reaction rate acceleration (often 10^4-fold), turnover of the substrate molecules into products (about 1000 molecules processed per second), and the re-use of same enzyme molecules for further reaction are the hallmarks of enzymatic processes. Figure 1 highlights some features of enzymology. An appreciation of the phenomenon may be had when we realize that almost all reactions of importance are enzyme catalysed, and that a whole life form, such as the lowly microbe, can be generated from its parent within twenty minutes — a dramatic and harmonic orchestration of the tens of thousands of chemical reactions involved in the effort ! Catalysts used in chemical laboratories and industry fall far short of this type of efficiency; it is thus no wonder that enzymes are being increasingly used in industrial processes. One aspect of biotechnology and bioengineering is to attempt to immobilize enzymes on solids and surfaces and use them repeatedly, as also to attempt to modify, design and generate new enzyme molecules for custom-made processes and reactions. Recombinant DNA methods and biotechnology have advanced now to a level that enables us to chemically

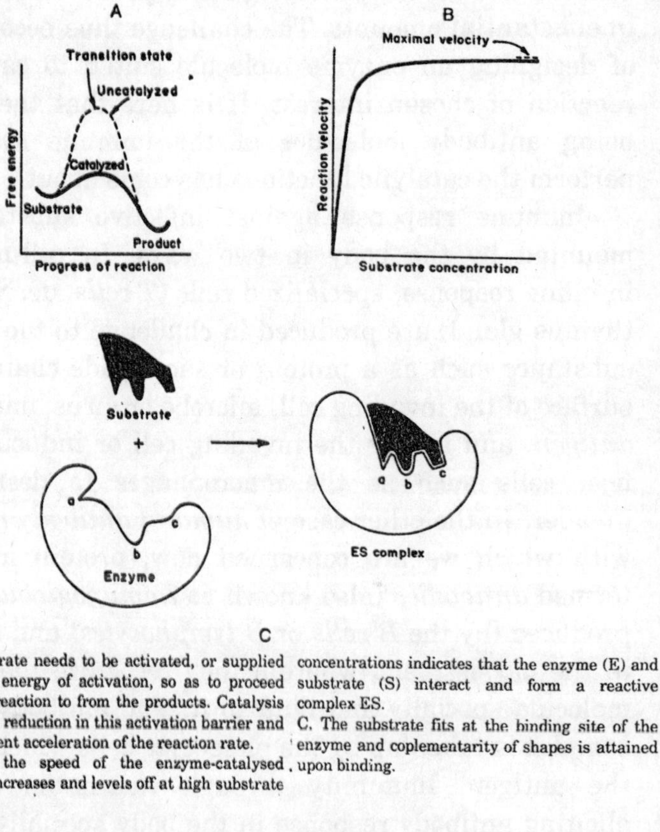

Fig.1 A. Asubstrate needs to be activated, or supplied with free energy of activation, so as to proceed with the reaction to from the products. Catalysis involved a reduction in this activation barrier and a consequent acceleration of the reaction rate.
B. That the speed of the enzyme-catalysed reaction increases and levels off at high substrate concentrations indicates that the enzyme (E) and substrate (S) interact and form a reactive complex ES.
C. The substrate fits at the binding site of the enzyme and coplementarity of shapes is attained upon binding.

synthesize a DNA molecule which contains the information to synthesize the chosen enzyme molecule, incorporate this DNA into the genetic material of a microbe and let the micro-organism produce the enzyme in substantial amounts. The challenge thus becomes one of designing an enzyme molecule suited to catalyze a reaction of chosen interest. It is here that the idea of using antibody molecules of the immune system to perform the catalytic functions has come about.

Immune response against infective substances is mounted by the body in two ways. In cell-mediated immune response, specialized cells (*T cells*, derived from thymus gland) are produced in challenge to the 'foreign' substance such as a protein or saccharide chain on the surface of the invading cell, microbe or virus, namely the *antigen*, and killing the invading cell or inducing other host cells such as the macrophages to destroy the invader. In the other case of *humoral antibody response*, with which we are concerned now, protein molecules termed *antibodies* (also known as *immunoglobulins*) are produced (by the *B cells* or *B lymphocytes*) and circulate in the bloodstream, wherein they recognize the antigen molecule specially and bind to it. The antibody-antigen complex is taken up by the phagocyte cells which destroy the antigen. Immunity through vaccination involves eliciting antibody response in the body specially against the vaccinated antigen (usually dead or very highly attenuated populations of the infecting cells or their surface macro-molecules) and activating and sustaining this mode of defence for years in the body.

The paritcular points of interest to us here are :(i) such cells are made by the billions in the body (about 10^{12} lymphocytes in man), (ii) they can be activated by injecting an antigenic substance in the free or in protein-bound form (called a *hapten*) into experimental animals (mice, rats, goats or man), (iii) antibodies can be made against millions of different kinds of antigenic molecules and (iv) the recognition of the antigen by the antibody is an inter-molecular interactive process remarkably similar to that of the enzyme towards its substrate. Even more interestingly, an antigen stimulates many different clones of lymphocytes, where each *clone* is a population of identical cells derived by repeated division from a single ancestor cell and thus produces an antibody with the same antigen specificity. This specificity is related in the molecular level to the *epitope* or the shape of the exposed surfaces of the cell surface antigen molecule of the foreign system. There could be various antigenic determinants or molecular subsurfaces that are recognised by different but closely related antibodies. Figure 2 shows some of the molecular schematics of the immune response, from where the similarity of the process with enzyme-substrate can be seen to be striking. In order to obtain an antibody that is highly specific to a given structural feature or determinant of an antigen, one would need to identify and choose those antibodies that are produced from that specific clone which does this. This process is known as the production of *monoclonal antibodies*.

Figure 3 illustrates the method used to produce monoclonal antibodies, involving the culturing *in vitro* of the antibody-producing B-lymphocytes from the spleen of an animal which has been immunised previously with the chosen antigenic molecule. Since such spleen cells will die after a few days in culture, they are "immortalized" by fusing them with a mutant *myeloma* cell line derived from a tumor of B-lymphocytes. The resulting cells, termed *hybridomas*, produce the desired

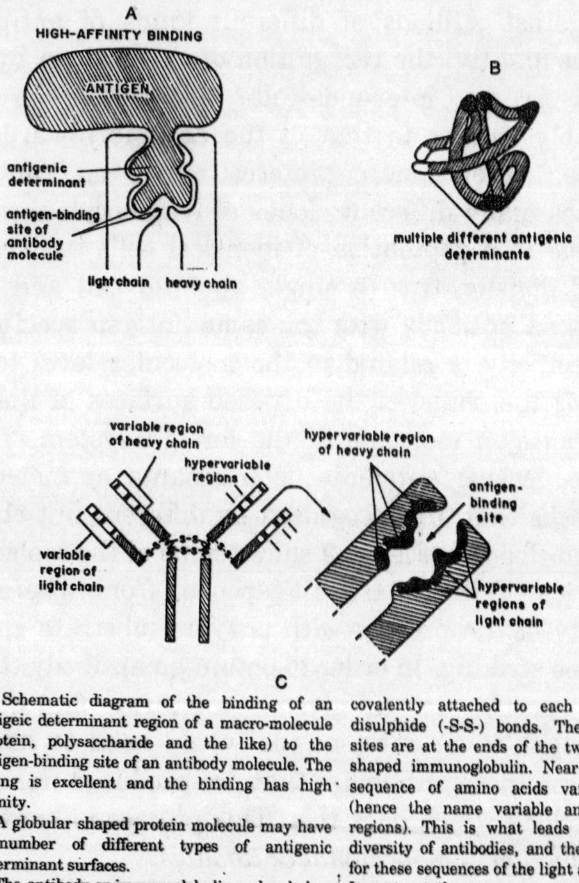

Fig.2 A. Schematic diagram of the binding of an antigeic determinant region of a macro-molecule (protein, polysaccharide and the like) to the antigen-binding site of an antibody molecule. The fitting is excellent and the binding has high affinity.

B. A globular shaped protein molecule may have a number of different types of antigenic determinant surfaces.

C. The antibody or immunoglobulin molecule has two identical light polypeptide chains (220 amino acids long) and two identical heavy chains (440 residues long). The light and heavy chains are covalently attached to each other by many disulphide (-S-S-) bonds. The antigen-binding sites are at the ends of the two stems of the Y shaped immunoglobulin. Near this region, the sequence of amino acids varies considerably (hence the name variable and hyper-variable regions). This is what leads to the enormous diversity of antibodies, and the genes that code for these sequences of the light and heavy chains. In contrast, the constant sequence regions, shown as shaded areas, have a set number of genes coding for them.

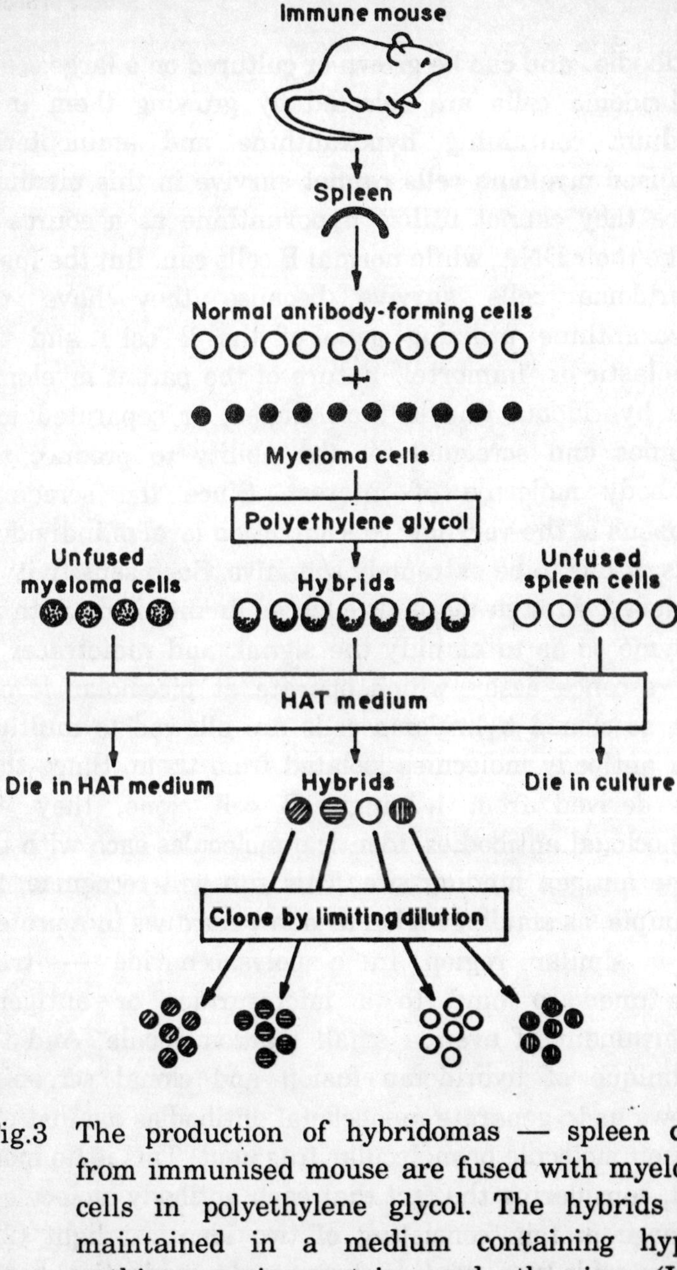

Fig.3 The production of hybridomas — spleen cells from immunised mouse are fused with myeloma cells in polyethylene glycol. The hybrids are maintained in a medium containing hypoxanthine, amino-protein and thymine (HAT medium), so only the hybrids survive. These are screened for antibody of interest and then cloned.

antibodies and can be grown or cultured on a large scale. Hybridoma cells are selected by growing them in a medium containing hypoxanthine and aminopterin. Unfused myeloma cells cannot survive in this medium, since they cannot utilize hypoxanthine as a source to make their DNA, while normal B cells can. But the fused hybridoma cells survive because they have the hypoxanthine utilizing gene of the B cells and the neoplastic or "immortal" nature of the parent myeloma. The hybridoma line is then cloned, or separated into colonies and screened for the ability to produce the antibody molecule of interest. Since the screening happens at the very low concentration level of individual cells, it has to be extremely sensitive. Such sensitivity is achieved through methods such as linking it up with an enzyme so as to amplify the signal, and radiotracer or fluorescence assay which operate at picomolar levels. The so cloned hybridoma cells are allowed to multiply and antibody molecules isolated from them. Since they are derived from a single B cell clone, they are monoclonal antibodies, identical molecules each with the same antigen binding site. This site will recognize, for example, as small a region as a few residues in a protein or a similar region in a polysaccharide — truly fine-tuned to bind to a microsurface or antigenic determinant of even a small macromolecule. And the technique of hybridoma fusion and clonal screening allows us to generate monoclonal antibodies against any chosen molecule or molecular fragment. This is no mean feat, considering the fact that each antibody molecule is a large protein consisting of two identical light (220 amino acids long) peptide chains and two identical heavy chains (440 amino acid residues), with very high variation in the amino acid sequence in the binding site regions (see Figure 2).

While monoclonal antibodies can be raised against any molecule or even a molecular fragment, enzymes are not available in nature to catalyse any reaction of our interest. And it would be of great technological value or even therapeutic utility, to be able to tailor-make in the laboratory enzymes that catalyse specific target reactions, e.g., to selectively hydrolyse protein or sugar coats of viruses or cancer cells, literally "*designer enzymes off the bottle*". The bisociation or fusion of the antibody idea and the enzyme idea to produce the concept of an *antibody enzyme* or "*abzyme*" or a "*catalytic antibody*" has not been too late in coming. Professor Peter G. Schultz and colleagues at the University of Calilfornia at Berkeley raised an antibody against compound I in Figure 4a. The idea pursued here was that the antibody molecule generated would have a negatively charged carboxylate group at a position complementary to the positive trimethylammonium moiety of the hapten, and this charge complementarity idea was realized in practice. Now, when this tailor-made antibody was mixed with compound II of Figure 4a, the carboxylate group of the antibody could catalyse the removal of HF, thus making the antibody an abzyme.

It has also been noted that while antibodies bind to the ground state of the antigen, enzymes selectively bind to the transition state (the top of the hill point in Figure 1) of the reaction. Therefore, an antibody that can be made to bind the transition state can be expected to have enzyme-like catalytic ability. Since the transition state is a very short-lived entity, not easily captured and worked with in the laboratory, chemists synthesize and work

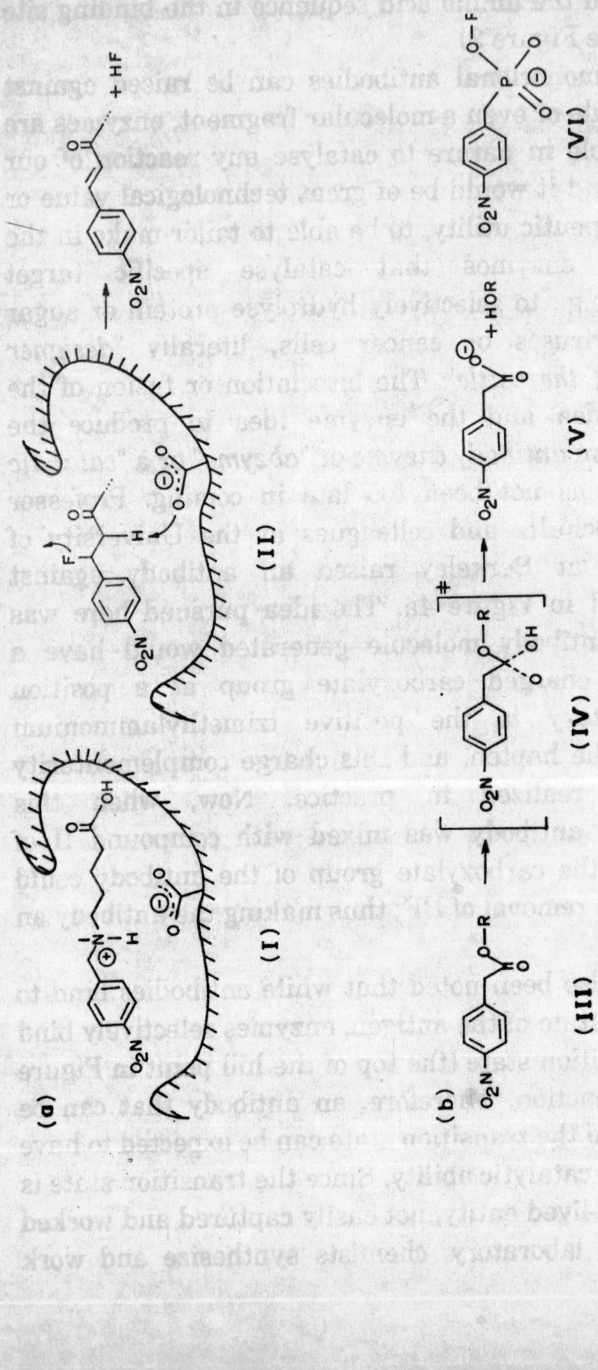

Fig. 4 The antibody raised compound I hapten is shown as the larger surface binding to the antigen. Charge complementarity has led to the generation of the carboxylate group in the antibody contiguous to the positive ammonium group of the antigen. When this antibody binds compound II, the acid group catalyse the unsaturation reaction.

with compounds that are analogues of the transition state. Structure IV in Figure 4(b) is the transition state of compound III, produced during the reaction of III to produce the final product V. The phosphate compound VI is the transition state analogue of IV.

Professor Richard A. Lerner of the Research Institute of the Scripps Clinic and his associates exploited this point and have obtained antibodies against transition state analogues such as VI. The anti-VI antibody so produced displayed the expected catalytic ability in the hydrolysis of III to form V.

The foregoing discussion has involved the use of hybridoma technology to generate monoclonal antibodies. While this method is adequate with rat or mouse lymphocytes, culturing human cells has posed considerable problems. It is in this connection that an important breakthrough has been achieved by the Scripps Clinic groups. Rather than using hybridoma methods, they have utilized the *gene cloning* method. Figure 5 shows a schematic illustration of this remarkable approach. In the first step, the transition state analogue molecule (NPN) corresponding to the reaction that needs to be catalysed is made and injected into a mouse, so as to elicit the anti-NPN antibody. Next, the messenger RNA molecules (intermediaries between the genes or DNA, that code for the antibodies and the Protein antibodies themselves) are isolated from the spleen of the immunised mouse. It is easier to isolate the RNA and to make the genes or DNA from the mRNA than to hunt for the antibody genes from the 'haystack' of the entire chromosomes of the mouse. The DNA or gene sequences so obtained from mRNA can be made in substantial amounts, using the recently developed powerful method called the polymerase chain reaction or

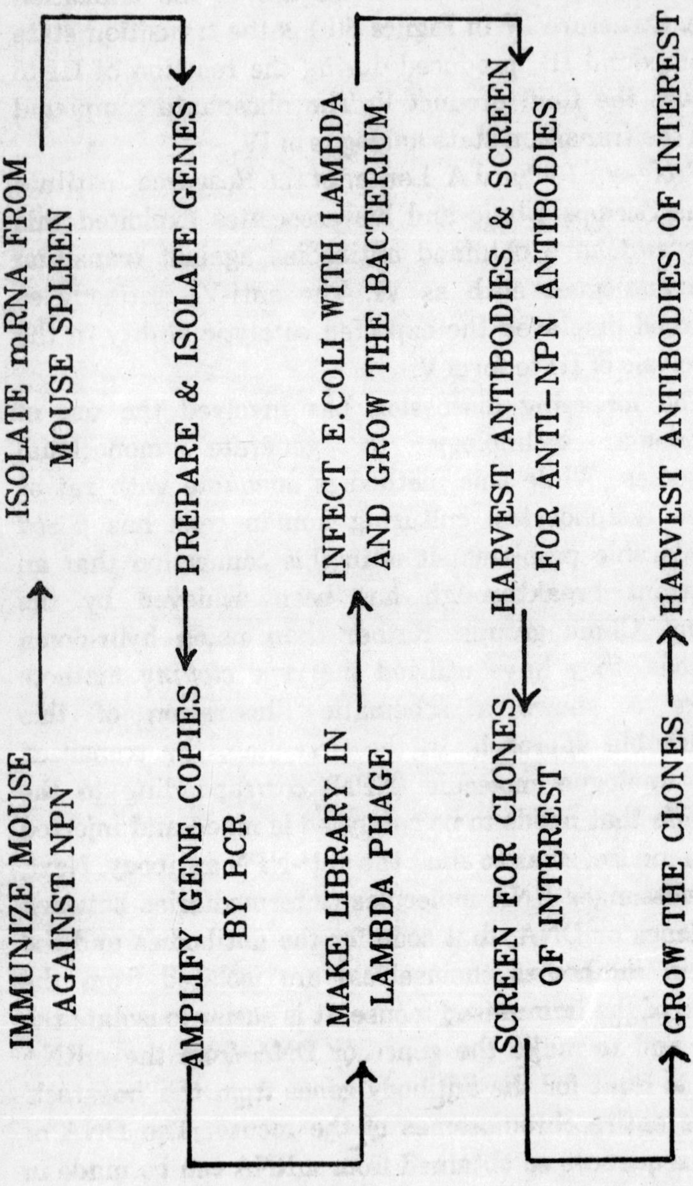
IMMUNIZE MOUSE AGAINST NPN
ISOLATE mRNA FROM MOUSE SPLEEN
PREPARE & ISOLATE GENES
AMPLIFY GENE COPIES BY PCR
MAKE LIBRARY IN LAMBDA PHAGE
INFECT E.COLI WITH LAMBDA AND GROW THE BACTERIUM
HARVEST ANTI-BODIES & SCREEN FOR ANTI-NPN ANTIBODIES
SCREEN FOR CLONES OF INTEREST
GROW THE CLONES
HARVEST ANTIBODIES OF INTEREST

PCR. The DNA so produced is recombined or introduced into the bacteriophage called *Lambda*, a virus that infects bacteria such as *Escherichia coli* (E. Coli)and integrates its own DNA into that of the host bacterium. Lambda thus serves essentially as a vehicle or a transfer device to introduce the mouse antibody genes into *E. coli*, which can now be cultured in large amounts. From the so grown *E. coli* population, Lerner and associates screened for the anti-NPN antibody by a special screening method and isolated the clones of interest. These clones were further grown and harvested for the anti-NPN antibody molecules.

Lerner's method opens the door for the creation in the near future, of a genetic library which would contain a very large (10^{12}) number of different antibody fragments. This library would also make the rapid production and analysis of catalytic antibodies easier — *truly a case of monoclonal antibodies and abzymes off the bottle*!

To Be or not to Be: Challenge of Tomorrow and of Today

M.P. Parameswaran

The development of science and technology — man's understanding of the laws of nature and of the capacity to transform it — has made it possible for every human being in this world to satisfy his basic biological needs like food, clothing, shelter and health care. Not only that. All can be provided with services to satisfy their cultural requirements like education and recreation, too. A truly human life, not an animal like existence, is possible for all. There can be no dispute over this. But the reality is that nearly half of the humanity live under sub-human conditions. The reason for this is, perhaps, that human beings have not learned to organize themselves to create conditions in which it is neither necessary nor possible for a small group to become ferocious animals brandishing nuclear weapons and star war threats and keeping the majority under subhuman conditions. Attempts to reorganize human society have been going on for long, the most successful being those made in the socialist countries. But recent developments in some such countries show that things are not as rosy as they have been supposed to be. If one looks closely into the happenings in these countries, one can see one factor common to all: absence of the participation of the people in running the affairs of the society. True, this is the case not only in these countries but also in the so-called 'democracies' of the world, including the oft-quoted "largest democracy".

The call of the present epoch is for a participative democratic social order. This is not easy to achieve. Obviously, no sensible person will welcome an exploitative and oppressive social order. History cannot go backwards. The present happenings in the socialist block cannot take it back to the days of capitalism with attendant unemployment, poverty and inequality. Neither subjectively nor objectively does humanity want a world threatened by nuclear war and environmental degradation, by multimillionaires and mafias. When there is a possibility to build a heaven on this earth, why should it be converted into a hell? But to achieve this a new form of social management is required. It is no longer desirable to leave the affairs of the society in the hands of a few chosen leaders. Experience has shown it. The only alternative is effective democracy where all the members of the society take part creatively in the running of its affairs. This is easier said than done. The very development of man's ability to transform nature, to develop science and technology, has made effective democracy more difficult. Effective democracy at this stage of human development requires relative autonomies to local groups or, to put is in another form, larger local control on local man-nature relations.

This is not possible if only a selected few in the society control the forces of production. The entire community should be able to control these forces. They cannot be left to the 'leaders' and 'specialists' alone. This is what the present epoch demands. Either we rise to the occasion, or face the consequences, which may possibly be the total annihilation of life itself on this planet. This is the challenge before us today. It is not a challenge before the political leaders alone. It is more of a challenge before the scientific and technical community. It demands a higher stage in the development of science and technology, a stage which will make 'small' not only 'beautiful' but also 'powerful'. The historical impact of the development of science and technology has been one of agglomeration, of motion towards 'mega systems'. Small could not sustain themselves. They became powerless, helpless. This does not require much elaboration. A limited number of trans-national corporations today control almost the entire humanity. "Big is powerful, and hence beautiful" — this is the philosophy of the day. And what are the fruits of this philosophy?

Uprooting of larger and larger chunks of population, driving them to cancerous metropolitan formations, submitting them to various stresses arising from lack of drinking water, sanitation, shelter, privacy and above all, deprivation of the feeling of belonging, catastrophic degradation of micro and macro environments, utterly wasteful exploitation of precious and limited resources, destructive competition developing into a propensity to go to risk, even at the slightest provocation — these are some of the bitter fruits which humanity is forced to eat. This is not to deny the wonderful achievements with potentials to build a heaven on earth. The question before humanity is how to retain the sweet, to discard the bitter and actually build the heaven on earth. This is the challenge before humanity today.

Is it possible for them to produce the necessities and comforts of life without the necessity of leaving their homes and environment, without commuting daily a hundred kilometres or more, with its consequences on vehicles, roads, accidents, energy and pollution? Is it possible to increase agricultural productivity without increasing, in fact, even decreasing, the non-renewable energy inputs in the form of fertilizers and pesticides, without causing ecological disturbances and pollution? Is it possible to produce goods which humanity consumes today at much smaller scales, in widely dispersed locations without increasing, in fact, decreasing inputs of energy, materials and labour per unit output? Is it possible to arrest and reverse the tendency of increasing mass transportation of human beings and materials?

This is the greatest challenge before science and technology today. This is the challenge before the society today. This challenge cannot be met by scientists alone. Only a scientifically oriented and well-informed society can face this challenge. The emergence of such a society in India can have significant impact on the world. The People's Science Movement can play and is playing an important role in this. People's Science Movement is of recent origin. The emergence of it in India can be traced to the experiments and experiences of Kerala Sastra Sahitya Parishad (KSSP) and its links with other groups and organizations in India.

The KSSP was started in 1962 as a small group of science writers, about 40 in number, with the objective of enriching the Malayalam language with scientific literature. From 1962 to 1990 it has travelled a long way. Today, it is a major organization with about 50,000 members, who come, literally, from all walks of life. During the past two decades, the objectives of KSSP have become wider to encompass practically all human activities. It is active in the fields of education, health, environment, energy, agriculture, irrigation, planning, in fact, almost every field of social life. This transformation of KSSP from mere science popularization to activism took place through intense interaction with the people.

In 1974, KSSP adopted the slogan of "Science for Social Revolution". It characterized the Indian society broadly into two groups: a majority which is either continuously getting impoverished or facing the threat of impoverishment and a minority which is getting continuously enriched at their expense. Social revolution is characterized by a reversal of this process. Science should become a powerful weapon in the hands of the majority in their fight against impoverishment. An important characteristic of People's Science Movement is its partisanship with the cause of the majority.

Several science groups have been emerging in various parts of India. Time was ripe to bring together all these groups to share experience, dream together, think together, plan together and act together. All India workshops were organized in Kerala during 1978, 1981, 1983 and 1985. These culminated in the *Bharat Jan Vigyan Jatha in* 1987 and the formation of All - India People's Science Network (AIPSN). Since then the AIPSN has been organizing every year an All - India People's Science Congress (1988 - Cannanore, Kerala; 1989 - Calcutta; and 1990 - Bangalore). These People's Science Congresses addressed themselves to problems of environment, health, development, education and self-reliance.

But these problems will remain unsolved so long as India remains illiterate. Literacy has to be on immediate agenda before the People's Science Movement. There are more than 400 million illiterates in India. More than half of them fall within the productive age group of 15-45 years. Experience has shown that about 20 million volunteers will be required to make 200 million people literate. Obviously this is not a small number. However, if this many number of volunteers can be raised over a period of three to four years and if 200 million people become involved in the program as learners, it will be an expression of a national will never hitherto demonstrated. The sheer magnitude of this will would cut across all the caste, religious and linguistic barriers. The experiment which would, it is expected, lead to the emergence of such a people's upsurge is known as *Bharat Gyan Vigyan Jatha.* Essentially, it involves the setting up of 60,000 - 70,000 local organizing committees, with a total membership of one million or more, drawn from about 5 lakh villages, and also the recruitment of one million volunteers for literacy work. This will bring forth people 'possessed' with a learning hysteria. Literacy work would become a people's festival.

Not only literacy work, but also the fight against conditions which perpetuated illiteracy would assume a new dimension. It will help creation of a genuine scientific temperament among the people. It will help the struggle to bridge the ridiculous gap between the possibilities offered by the breath-taking strides in science and technology on the one hand and the pathetic reality with more than half the humanity living under extremely sub-human conditions, on the other.

It will help the people to take their destiny in their own hands, help them to take steps towards a truly democratic society. This demands the development of certain skills and capabilities in a massive way. It cannot rely on a 'few leaders' educated in the selected institutions of excellence and blessed with positions of authority. People should be able to plan for themselves, decide what is good and what is bad for them. They should be able to understand the pros and cons of various developmental debates taking place around them. This is doubly important today, because many of these can have far-reaching consequences. They should be able to understand this and also should be capable of making cost-benefit analysis in their own way. This will be accepting people's right to know, backed up by 'ability to learn'. It should also be, supported by freedom to use the knowledge.

A major experiment has been initiated in Kerala to evolve a methodology for this. Research institutions like Centre for Earth Science Studies, Centre for Development Studies, Centre for Water Resource Development and Management and also departments like Agricultural Department, Irrigation Department, State Planning Board, and State Land Use Board together with the *Panchayat* Association and the KSSP are involved in this programme. About 1000 persons, including people's representatives, local body executives and volunteers from KSSP will be trained rigorously and with the help and active participation of volunteers from the locality, detailed *panchayat* /town level resource bases will be prepared. This will form the basis for future development planning. Detailed *panchayat* level development plans too will be prepared. A system of presenting development plans before the public for debate and criticism too is being worked out. It is a long and tedious process. There is no short-cut to participative and creative democracy. Scientists and technologists have an important role to play in this. Without their help the society cannot move from passive democracy to participative democracy. It is important for science and scientists too. Today they are being pushed more and more into realms of irrelevance. Unless surrounded by the medium of creative and democratic society, science and scientific community in India will become baseless and even cancerous. How to avert this?

This is the second challenge before scientists and technologists in India. Can they learn to seek inspiration from the aspirations of the suffering millions of the country? Can they learn to attain professional satisfaction, not through publication abroad or through peer appreciation, but through work relevant to the majority who are continuously getting impoverished? Can they learn to understand the language of the people, converse with them and understand them? This is the challenge before the scientific community in India today.

People's Science Movement can be of help to them. It is a people's movement. It is a democratic movement. Above all, it is a science movement. It is a movement that would make science relevant to the people, make scientists useful and dear to the people. Accepting this challenge, scientists will find themselves in a new company, a much more enjoyable company. The scientific challenges they would have to face will be more difficult than that faced in developing a Grand Unified Theory, or unravelling the complexity of human brain. It is a challenge to make the small powerful. It is a challenge to train millions and millions to a level of capability unthinkable today. It is a challenge to build real democracy. The People's Science Movement offers to scientists and technologists of India a platform to take up this challenge.

Intellectual Property Rights and Technical Change in Follower Countries

Amiya Kumar Bagchi

Defining property rights in knowledge, information, or reputation is a capitalist innovation. The practice started in the Italian city states : it seems to have been first formalized in a Venetian law of 1474. From the Italian city states, the practice spread to other countries of western Europe. In England during the reign of Elizabeth I, her minister, Lord Burghley, granted a series of patents with a view to encouraging foreign inventors to import their inventions and work them in England, and stimulating inventions by domestic producers. The royal power of creating monopolies was often abused, so by the Statute of Monopolies of 1624, restrictions were placed on the power of the monarch to create monopolies; but granting monopolies for the ownership and use of inventions was specifically left out of the restrictions[1].

England passed a series of Acts granting patents in inventions, and these were consolidated in an *Act of 1852*. A very similar law was passed by the British

Indian Government in 1856. This is the origin of patent legislation in India. The current law was passed in 1970, though not enforced fully until 1972.

Intellectual property rights also cover trade-marks and copyright. As in the case of the patent legislation, laws relating to these types of 'Intellectual Property' date back to the days of British rule in India. The present article will be confined mainly to the issue of legislation for conferring ownership rights in inventions, that is, to patent legislation.

It is useful here to recall the rationale for patent Acts. They constitute essentially a compromise between the social costs of using an invention, once it has been made, and the private costs of making an invention. Once an invention has been made, the cost of using it is virtually zero, so that ideally speaking, there should not be any charge for using it. On the other hand, individuals and firms have to incur major costs for making the invention[2]. There have always been individuals who invent for the sheer joy of making new things; and there will always be such individuals. However, in a commercialized society, such people will find it difficult to carry on their activities unless they are paid for (Henry Cavendish, after all, was enormously rich and could afford his hobbies); and research is becoming increasingly costly. So patents in inventions are granted as a reward, so that firms and individuals are induced to carry on research. Another solution is, of course, for the government to subsidize research. That solution is also not without its pitfalls, since only the result can tell whether expenditure on a particular piece of research was worthwhile or not. Since there is no obvious or easy way of assessing the ability or monitoring the effort of the scientists and the technologists involved, much expenditure can easily go waste[3]. In actual practice, most countries allow a combination of public subsidy and private appropriation of research results in the form of patents or inventors' certificates.

A national patent system also embodies another kind of compromise. If the world were ruled by a supernational government, it would subsidize research in any country and it might also issue patents to any individual or firm in any country. In such a world capital, labour and information or knowledge would move freely from any point on earth to any other point. In actual fact, the human race is divided into states, with their own laws and regulations, and with numerous restrictions on the movement of people, capital and knowledge.

National patent legislation then is aimed at attracting inventions from abroad, and encouraging inventions at home. Attracting foreign inventions or encouraging national inventions does not mean simply obtaining the blueprints, but getting them to work. This is the rationale of provisions for compulsory licensing of patents under certain conditions. India has taken over the clauses relating to compulsory licensing from laws of Western European countries, including Britain. If we can always bear in mind the aims of patent legislation in the abstract and those of national patent legislation in the concrete, then we can pinpoint the deficiencies of our R&D set-ups, the deficiencies in the working of our patent legislation, and bad faith behind many of the new proposals floated by the U.S.A. and the O.E.C.D. countries. These latter seek to do away with the national character of our patent legislation in a world which remains ineluctably divided into states with strong barriers against the mobility of people, capital and information.

Before we take up some of these issues, let me stress that in today's world of science and technology, India is a follower country. Most of the great scientific discoveries have been made abroad, and will continue to be made abroad. More than 99 per cent of the technologies used in modern industries, transport and communication networks were invented abroad, and transmitted to this country. Our aim in most fields of science, and even more importantly, of technology, must be to learn, absorb, adapt and catch up.

A few years back, I had put forward as an approximate criterion for technological self-reliance for the follower countries that the latter should be in position to import, adapt and absorb the technologies that are needed at the time and to develop those technologies for which need is felt but which cannot be imported from abroad (because such technologies have not been invented anywhere else, or because their import imposes prohibitive political and economic costs, or because, although available abroad, the country cannot obtain it either for love or for money)[4]. When we discuss policy choices in the area of science and technology, we must try to safeguard this freedom of choice — as to whether to import or not, borrow or not, locally develop or not — in every area, and must choose the best path from the point of view of our collective national advantage. This advantage will often lie in selective import and adaptation rather than an attempt to develop all the processes on our own.

Many of our scientists will, of course, work in frontier areas of science. But it is an illusion to imagine that in most areas we can seriously aim at making revolutionary innovations. This is especially true in the field of technology. To be conscious of our national needs is to abjure any false pride and arrogance and recognize that we need to learn as much from the foreigners as we can. The most outstanding example of successful learning on a national scale in modern times is Japan. Before the 1960s, there were few areas in which the Japanese had made any revolutionary discoveries. At least down to the middle of the 1970s, Japanese technological balance of payments, that is, the difference between the fees or royalties obtained from technology exports by her and those she paid for technology imports was a large negative figure. But Japan had already emerged as a top manufacturing nation of the world by the 1970s. By now she is either the second or the third largest producer of manufactures in the world as a whole.

We have to anchor our policy firmly on the recognition that we will continue to be a follower country for some time but aim to be a successful follower. Paradoxically enough, in many areas this requires a deepening of the national orientation of our technology policy, rather than a slavish imitation of practices of advanced countries — or worse, a slavish obedience to the directives being issued from the OECD countries.

A sensible technology policy must be seen as an integral component of an economic policy aimed at sustained growth, with at least no increase in the degree of inequality of distribution of economic resources. For a start, proper implantation of many technologies requires adequate diffusion. As technologies become further and further elaborated, the number of components or processes required to develop the final product also increases enormously. A country can be said to have really absorbed a technology properly when it has numerous technologies specializing in each step of the process of production. Development of such skills requires diffusion of education and proliferation of organizations which can utilize the skills of those technologists. Moreover, every technology requires to be adapted locally in however minor a degree. A wide diffusion of technologies in turn leads to their improvement with mechanics, artisans and engineers making minor innovations as they learn the techniques and help to combine them in new ways.

All of these processes are facilitated by education. Universal education has been a pre-requisite for sustained industrial growth in every advanced manufacturing nation. The district of Ernakulam in Kerala has recently celebrated the eradication of illiteracy. Technological self-reliance will require the replication of the Ernakulam story in every district of India. Basic scientific and technical education can only be imparted in the mother tongue of the practitioners. In how many states in India is engineering or scientific education even up to a B.Sc. Pass or Engineering Diploma level imparted in the local language? Even where there are textbooks in the local language, they are rarely written by active scientists and engineers. And few new problems get written up in the local language. Again the contrast with Japan is devastatingly obvious, where until quite recently, most of the research was being written up in Japanese. So an eagerness to learn is not antithetical to a sense of national identity but is a strong reinforcing element. (Similarly in the sixteenth and seventeenth centuries, Englishmen learned from the Italians and the Dutch but learned through and in their own language).

Another area where we should learn how to learn is that of the design of our organizations. Both the colonial legacy and the Brahmanical tradition left us with organizations which are too hierarchical. Again, the Japanese and, increasingly, the European experience shows that a flattening of the hierarchical order, and a network of horizontal linkages at the actual levels of production are good for dissemination of learning and generation of new ideas through teamwork.

Let me pass on to the subject at issue — the sort of intellectual property rights we should have, and the use we should make of whatever systems we have adopted. For the moment, let us assume that the compromise we have adopted is a patent system — but a national patent system, not a sham international system where all the controls are in the hands of a group of advanced capitalist systems. The 1970 *Patents Act* constitutes a national patent system. The tests we should apply to see whether that Act has worked well or not are: (a) whether the total number of patents has increased over time; (b) whether Indian firms and individuals have substantially increased their share in the total number of patents; (c) whether Indian firms have been able to take out a larger number of patents in foreign countries; and (d) whether there has been a trend towards a much larger number of foreign patents being licensed to Indian firms.

I have not looked at the latest patent statistics. But I had looked at the data up to the beginning of the 1980s. Going by these data, the only aspect in which it can be said that the potential of the *Indian Patents Act* of 1970 was remotely realized was an increase in the percentage of the total number of patents taken out by Indian firms and individuals[5]. But this is small consolation in view of the fact that the total number of patents registered in

India, which was only a fraction of the total number of patents taken out in the USA or Japan to start with, came down sharply in the 1970s. When the foreigners stopped taking out patents for fear of disclosure of critical information and because of the curtailment of the duration of validity of patents in the areas of chemicals, drugs and pharmaceuticals, this in fact provided an opportunity to Indians to scour the world for information on new technologies and get in first with their own applications. There is no evidence that, except to some extent in the field of drugs and pharmaceuticals[6], Indian firms or individuals really seized this opportunity.

One index of the utilization by Indians of the information available in patents that have been filed and sealed could have been the collection, filing and scrutiny of such information in a systematic manner. For an idea of the scale of effort made in studying patent specifications abroad, it may be mentioned that the U.S. Patent and Trade Mark Office (almost certainly the biggest technological library in the world) by 1980 contained more than 22 million documents, and documents at the rate of 600,000 a year were being added to the collection[7]. By contrast, a patent information system was just being organized in India in the early 1980s, and as far as I know, much of the information collected has not been classified yet. This is not only a default on the part of the government, it is also a lapse on the part of the business community. If the latter had been keen on utilizing the information available the world over in the form of patent specifications, then by now a more systematic drive for collection and classification of this information would have emerged in public view.

There is another aspect of the utilization of a patent system in which again Indian firms have performed quite poorly. This is the extent to which Indian firms have taken out licences of foreign patents. The so-called foreign collaboration agreements rarely contain any patent licensing provision. What they provide is a renting or leasing (for a strictly limited period) of unpatented knowhow; the blueprints and designs of such knowhow have generally to be returned to the foreign firm at the conclusion of the agreement. Such foreign collaboration agreements have largely defeated the provisions for curtailing the lifetime of valid patents, because through repeated renewals (in different forms) foreign firms can retain control over the technology leased to Indian firms even when they have no real proprietary monopoly in such technology. The provision for compulsory licensing of patents not worked in India has also been virtually a dead letter for the reasons stated above and because of various legal problems.

Generally speaking, the degree of cross-licensing of patents to firms in a particular country by owners of patents worldwide varies directly with the level of development of a country[8]. It is not terribly surprising then that India does badly by this criterion. But I have a hunch, that if we take account of the size of India's manufacturing sector, and not just of her income per capita, she does badly by the average standard of the so-called newly industrializing countries. The ability of a country to exploit the enormous resources of knowledge available globally depends not only on its average income but also on the degree of dynamism, the extent of willingness to commit resources for long term investment, and the degree of competitiveness displayed by its industry.

The awareness of the dimension of international competitiveness of technology development is quite limited in official circles. For example, the *Annual Reports* of the Ministry of Science and Technology contain no data on patenting by Indians either in India or abroad. The figures of R&D released by it are quite misleading, for as is well-known, much of the so-called R&D expenditure is either spent on quality control or on public relations exercises. The figures of qualified scientific and technological manpower are equally misleading, given the enormous variation in quality as between different institutions and the purely ascriptive character of the certification carried out by many institutions.

But while the utilization of our patent and R&D system is quite defective, changing it towards the direction of freer trade would be a perverse solution to the problem. It would be akin to saying that since I don't like sick patients, it is best to kill them off. What the U.S.A. and other advanced market economies are attempting is to take advantage of their initial lead, and see to it that all the countries trying to catch up with them never have a chance to do so. In the process, they are causing enormous damage to a highly recessionary world economy; but their philosophy is that they would rather be dominant partners in a shrunken world economy, than allow new countries to come up as serious competitors in an expanding world economy.

As far as intellectual property rights are concerned, the US and OECD lobby are conducting their campaign on two major fronts. The first is to bring all the Less Developed Countries (LDCs) under the Paris Convention and whittle down the national privileges of every nation to a vanishing point. Countries whose firms and organizations have no capacity for carrying out a worldwide search for priority are told that they are free to conduct a search and claim priority for any of their patentable inventions if they can. But reciprocally all firms in any other country (and these can only mean trans-national corporations based in the OECD countries) can file a patent application in any of the Paris Convention countries and claim priority in any other country. The latter will by this device practically eliminate the possibility of any but a few LDC firms or individuals ever successfully patenting an invention: the latter cannot match the information networks and the search powers of the TNCs operating worldwide.

The second and an even more insidious attack is being mounted under the Uruguay Round of talks for widening the scope of the GATT[9]. Here, the attempt is to define so-called trade-related intellectual property rights. One particular way in which this is being defined is to say that a patent has been used when the patentee or a licensee has used it to produce a product anywhere in the world. In that case, if the patentee or an agent of his imports a product using the patented invention, then the patentee will be said to have used that patent in the importing country also. Going even further, it is suggested that any country which is a signatory to the new GATT conventions must allow the free import of those products (barring perhaps tariffs of a fairly low order — to be approved by the other signatories).

Such a step will, of course, put further barriers against the industrialization of the LDCs. Apart from the eventual effect in slowing down global economic growth and deepening the poverty in the poorer two-thirds of the world population, the bad faith of these proposals is also obvious. For a start, free trade includes only free trade in commodities, and one-way flow of capital (mostly between the OECD countries and from the LDCs to the OECD countries), and it totally excludes free trade in labour. Why does the USA permit the entry of skilled scientists and technologists and moneyed businessmen from all over the world and severely restrict the entry of the not-so-skilled workers and peasants? Do commodities not embody the labour of such people also? Secondly, all these negotiations are taking place against the backdrop of a huge debt overhang of the LDCs which has by now surpassed the mind-boggling figure of 1300 billion dollars. These countries are for ever being told to tighten their belts, exercise fiscal discipline and pay up, while the U.S. budget deficits climb from one astronomical height to another every year.

If we want to fight the blackmail of the OECD countries, we must see that we never get into the situation of a Brazil, Mexico, Argentina or Philippines, however tempting import-dependent new technology and consumer goods appear to be to our goggle-eyed rich. But we must also try to utilize our existing patent and R&D system far more effectively than we have done in the past and study possible ways in which we can change it to increase our own advantages. We could look at the Japanese patent system as an example from which to pick bits and pieces conferring on us a national advantage[10]. But that issue must be thoroughly debated between scientists, technologists, lawyers and economists committed to improving the living standards of the majority of Indians rather than of only the few whose eyes are glued to the dazzling spectacle in the affluent market economies.

Notes & References

1. Christine Macleod : Inventing the Industrial Revolution : The English Patent System 1660-1800, Cambridge: Cambridge University Press, 1988, Chapter 1; and R. Cooter and T. Ulen : *Law and Economics,* Glenville, Illinois, Scote, Foresman and Co., 1988, pp. 135.
2. The basic economic theory behind, the conflict between social and private costs and benefits of invention is laid out in K.J.Arrow: 'Economic Welfare and the Allocation of Resources for Invention, in *The Rate of Direction of Inventive Activity* (Princeton, N.J. : Princeton University Press for National Bureau of Economic Research, 1962).
3. P. Dasgupta : 'The Economic Theory of Technology Policy : An Introduction', in P. Dasgupta and P. Stoneman (eds.) : *Economic Policy and Technological Performance,* Cambridge : Cambridge University Press, 1987, pp. 7-23.
4. A.K.Bagchi : 'The Pursuit of Self-reliance in a World of Flux', (G.L.Mehta Memorial Lecture for 1986), in *Facets of India's Development* (Bombay, Industrial Credit & Investment Corporation of India, 1986), pp. 111.
5. A.K.Bagchi, P.Banerjee and U.K.Bhattacharya : 'Indian Patents Act and its Relation to Technological Development in India : A Preliminary Investigation', *Economic and Political Weekly,* XIX(7), 16 February 1984, pp. 287-304.
6. B.K.Kealya : 'Chemical Based Industries -- Drugs and Pharmaceutical and Pesticides : Foreign Pressure for Changing the Indian Patents Act, 1970', in *Occasional Papers,* Vol. XIV of National

Conference of Scientists on Science, Technology and Patent Laws, held at Indian National Science Academy on 4 December 1989 (New Delhi, National Working Group on Patent Laws, 1989).

7. D.A.Burge : *Patent and Trademark : Tactics and Practice*, New York: John Wiley & Sons, 1980, pp. 13-14.
8. W.L.Bosworth : 'The Transfer of U.S. Technology Abroad', Research Policy, Vol.9, October 1980; and F.J.Contractor : 'The composition of technology fees and arrangements as a function of economic development of technology by recipient nations', *Journal of International Business Studies,* 11(3), Winter 1980.
9. Deepak Nayyar : 'Some Reflections on the Uruguay Round and Trade in Service', *Journal of World Trade Law*, 22(5) October 1988; C.Niranjana Rao : 'Recent developments in the international patent system', *Economic and Political Weekly,* Vol. XXIV, Nos. 51 & 52, December 24-31, 1989.
10. See, for an exposition of the peculiarities of the Japanese patent system, A.Wineberg : 'The Japanese Patent System : A Non-tariff Barrier to Foreign Business', *Journal of World Trade Law, 22(1)*, February 1988.

Patent Laws : The Indian Experience

Nitya Anand

The subject of intellectual property rights (IPR's) and their protection, particularly of patent rights, has become a central issue in scientific, technological and economic development, and in scientific cooperation between industrialized and developing countries. Historically, the patent system evolved to protect property rights in innovations which were products of manufacture. Agricultural processes and products were traditionally considered to be outside the ambit of patent laws. Living things were also excluded from patentability, as these were regarded as products of nature rather than of manufacture, and as such were considered to be the common heritage of mankind, which should be available freely to everyone. This situation, however, has drastically changed in the last two decades, particularly as a result of the technological revolution ushered in by recent developments in biotechnology which promise big financial returns through commercialization of living systems or their products.

As living forms and biotechnological products and processes are now moving to the centre of the industrial scene in developing countries, pressures are mounting to extend the scope of patent protection to them.

There is a feeling that the subject of patents is esoteric and highly technical and is best left to experts. However, in view of their direct impact on many spheres of everyday life, there is need for a better appreciation of the patent laws by the community. The patent laws affect the climate in which intellectual activity takes place, and thus have an impact on R&D activity; the patent laws have direct effect on industrial production, imports and availability of consumer goods and their prices and thus are of great societal concern. It is intended to examine in this article the evolution of the world patent system and its application in the emerging fields of biology and in this context discuss Indian problems, experience and expectations in this field.

Intellectual Property essentially includes the ownership rights on products of the mind, on intellectual work, which can be grouped as under:-

*** Copyrights**

- Artistic and literary works
- Informatics, databases
- Computer software, integrated circuits, etc.

*** Industrial property**

- Patents
- Inventors' certificates
- Industrial designs and utility models
- Trademarks, etc.

The purpose of recognizing intellectual property rights is to give some benefits to authors and inventors as an inducement to create and disseminate intellectual work, and thus to promote the progress of science, technology and useful arts. In order to benefit from copyright, an author has to publish the work, thus making it available to the public. In order to obtain a patent, the inventor has to reduce his ideas to useful applications; a patent is not granted merely for an idea.

Each of the areas listed above is important for societal progress. New dimensions have been added to each as a result of the technological advances that have been made in recent years, particularly in the areas of informatics and computer software, but the present article will be confined mainly to patent rights as this has become a subject of intense debate on account of the pressure being mounted on Indian government to change its patent laws.

Patents

A patent is a property right granted by the State to a patentee, excluding others, for a fixed period, from using the patented invention without the consent of the patentee. An invention without a patent is not necessarily a property right; however, a patent does not automatically confer a positive right to use an invention because freedom of use would be dependent on obtaining prior rights by licensing. A patent can be for a product or a process or both; a *product patent* provides exclusive rights to the product manufactured by any processes, while a *process patent* only protects one process, and the product can be manufactured by an alternative process and thus provides scope for innovation to develop alternative processes.

The purpose of a patent is to provide a system of monetary benefits to the inventor/ investor so as to promote inventive activity and commercialization of inventions. Another purpose is to disclose inventions at an early stage, thereby adding to the world pool of knowledge, which in turn would aid in new discoveries and inventions.

Credit for Innovation: Central to the concept of Patent System is the exclusive credit for the invention, given to a single individual or group of individuals, although such invention is not an isolated act/event but has drawn upon prior research and world pool of knowledge in the field. How much credit belongs to the individual (or the group) and how much they owe to the world scientific community! This questions the very basis of the patent system!

Patents, monopolies and restrictive practices

By the grant of a patent, the State provides to the patentee a statutory and exclusive right, which can lead to the creation of a monopoly. Monopolies affect a wide variety of interests -- those of the patentee, the licensee, the manufacturer, the consumer and the state. Monopolies can lead to restrictive practices, cartel formation, including output restriction, price hiking, limited availability and under-utilization of capacity and can stifle healthy competition. And if the patentee-monopolist is a foreign company in a developed country, these practices result in unnecessary imports by and/or outflow of royalties from a developing country, whose indigenous R&D is hindered. There is thus incompatibility between the existing patent system and the development process of developing countries. Therefore, patent laws are so designed as to safeguard the public and national interests, and provide a proper balance between conflicting interests.

Provisions generally incorporated in the patent systems of different countries to safeguard public and societal interests include the following:-

- Exclusion of certain categories of products from patentability such as medicines and food products;
- Reduced life of patents for products of great social relevance;
- Right of the State to use the patent;
- Compulsory licensing and licence of right to ensure working of the patent within a fixed period.
- Price fixation.
- Various measures which have been commonly used to overcome the anti-competition effect of patents include:
- Exclusion of product patents, and allowing only process patents, which encourage innovation, competitiveness and improvement of processes.
- Laws against mergers, cartel formation and take-overs.

Diversity of patent laws of different countries

The patent laws of a country thus attempt to strike a balance between private ownership, individual benefits and creation of monopolies on the one hand, and public interest and societal benefits on the other, between technological self-reliance and import-substitution and import of technology and goods. The provisions for achieving such a balance would vary, depending upon the developmental stage of a country, and, therefore, these provisions have varied from country to country and also changed over the course of time. Each country thus tries to evolve patent laws best suited to its national interests, in consonance with its social aspirations and goals, stage of development and political system. This has been the history of the development of the patent system in all the 1st world countries, which have also gone through the same stage of industrial and economic development through which we are now passing. How can the patent laws of a country which is starting on its industrialization process be the same as those of a highly industrialized country, when their technological and industrial bases and their resource positions are so vastly different? All EEC countries can now afford to have common patent laws, since their levels of technological development are comparable, but it is absurd to think of developing countries having the same laws. Moreover, uniformity in patent laws of EEC countries has come about only since the early 1980's, prior to this, patent laws varied considerably from country to country. Many countries have introduced changes in their laws, as they have gone along the process of industrialization resulting in changes in their developmental status and needs. Italy had no patent protection for drugs and pharmaceuticals between 1950 and 1980; chemical products were not patentable in the Federal Republic of Germany up to 1967, in the Nordic countries up to 1968, in Switzerland up to 1978, and in Spain these will be patentable only from 1992. In Japan, medicinal products could not be patented up to 1976 and in Canada they are patentable only if the process is also claimed. Thus, there is nothing like uniform patent laws for all countries. The patent laws of a country are essentially a bargain between conflicting interests and needs.

The Indian Patent Laws

The first Patent Act was promulgated in India by British rulers in 1856, primarily to reserve the vast potential market of India for the products of the then emerging manufacturers of Great Britain. The subsequent development of the patent laws and their impact leading to the Indian Patents Act, 1970 are summarized below :

1. Early History

- Patent Act No.VI of 1856: Exclusive privileges to inventors of new manufacture for 14 years;
- Patent Act No.XV of 1859: Earlier Act made more comprehensive and patterned on the British Patent Act of 1852; modified in 1872, 1873 and 1882.
- Indian Patent and Designs Act of 1911, designed on British Patent Act; separate patent administration, with a controller of patents, established.

2. Impact

- Indian markets became subservient to foreign industry, particularly British industry;
- Resulted in monopolistic control by foreign companies, and most of the industrial goods were import from abroad at exorbitant prices;
- Restrictions on industrial production hampered domestic industrialization
- Prevented import of essential goods from competitive sources in the international market, since patent protection included the right of import.

3. Post-Independence History

The Indian Government considering the restrictive effect of the Patent Act of 1911 on India's industrialization, appointed a committee to examine this issue.

The Patents and Designs Amendment Bill 1950, and the comprehensive Patent Bill, 1950, made some changes but, basic structure was left untouched;

Justice Bakhshi Tek Chand Committee, 1948, summed up as follows:

> "The Indian Patent system has failed in its main purpose namely, to stimulate invention among Indians and to encourage the development and exploitation of new inventions for industrial purposes in the country, so as to secure benefits thereof to the people".

Justice N.R. Ayyangar Committee, 1957-59, with wider terms of reference, produced a comprehensive report on the revision of the patent laws in 1959. It made the following major recommendations for revision of the patent laws, in addition to various suggestions on procedure:

- to define with precision which should be patentable and by rendering unpatentable certain inventions, the grant of patent to which will retard research or industrial progress or be detrimental to national health or well-being;
- to provide remedies for the evils which India, in common with other countries, experienced from foreign owned patents which are not worked in the country, but which are held either to block industries of that country or to secure a monopoly of importation;

- providing special provisions as regards the licensing of patents for inventions relating to food and medicines;
- Based on the recommendations of the Ayyangar Committee and extensive discussion, consultation, parliamentary debate and examination by two select committees of Parliament, the new *Indian Patents Act* was passed in 1970 and was enforced from 1972.

Indian Patents Act, 1970

The basic approach of this Act, as stated in Articles 83 (a) and (b) of the Patents Bill, based on public and national interests, is given below.

a. That patents are granted to encourage inventions and to secure that inventions are worked in India on a commercial scale and to the fullest extent that is reasonably practicable without undue delay; and
b. That they are not granted merely to enable patentees to enjoy a monopoly for the importation of the patented article.

Salient Features of the Patents Act, 1970

* Patentability

Product patent for all inventions, except for substances to be used as food, medicines, or drugs or relating to substances produced by chemical processes (including alloys, semi-conductors, optical glass and intermetallic compounds) for which only process patents.

Inventions relating to atomic energy, agriculture and horticulture and methods, of treatment of humans, animals and plants not patentable.

*** Term of Patent**

Normal term of all patents is 14 years, except for substances intended to be used as food, medicines or drugs for which 7 years from the date of application or 5 years from the date of sealing of patent, whichever is earlier.

*** Equal Rights**

No distinction between Indian nationals and foreign nationals regarding patent rights.

*** Compulsory Licensing**

If reasonable requirements of public interest not satisfied, any interested party can be licensed on application, after 3 years of less grant of patent.

*** Licence of Right**

If reasonable requirements of public interest about availability and at reasonable price, not served, government may endorse : 'Licenses of Right' for any patent; such endorsement to be gazetted by the Controller of Patents.

Process patents for food, medicine, drugs and chemical substances are automatically deemed to be endorsed with words 'License of Right after 3 years.'

The Indian Patents Act, 1970 has been a historic document, and in some ways a model to promote the development process in developing countries. It strikes a fair balance between private and public interests, between the interests of the developed and developing countries and promotes competitiveness in industry.

- Protects the interests of both the inventor and consumer in a balanced manner.

- Does not give patentee monopoly of importation at his own price and from his source.
- Gives priority to national interests over the interests of the foreign patentees.
- Helps development of processes in the field of pharmaceuticals, pesticides, and other chemicals leading to self-reliance.
- Process patent permits development of alternative technologies to suit Indian conditions; promotes research efforts and competitions in industry.

Indian Patents Act in the world context

There is nothing very special or unique in the laws of the Indian Patents Act, 1970; most of the provisions introduced in this act had precedents in the patent laws of developed countries during the early phase of industrialization and some of the these even now exist in their present patent laws. Some of the these provisions are examined below:

- *Compulsory Licensing*: countries which still have similar provisions as in Indian Patents Act are UK, Germany, France, Finland, Sweden and USSR. In fact model laws prepared by World Intellectual Property Organization (WIPO), Geneva, and its predecessor, BIPRI, include such a provision. *License of Right* provision exists in patent laws of Canada, U.K., France, Australia, China, Republic of Korea, Thailand and Philippines.
- *Only Process Patents for Chemical Products including drugs and medicines*. This is the most contentious provision and is examined at some length. The German Patents Law of 1877,

enacted when chemical industry revolution was under way in Europe, specially excluded product patents for chemical substances -- such patents would stultify the industry, *while challenge of new processes to develop would make the industry, more innovative and competitive which is the main objective of the patent laws*. Germany had excluded products from patentability till 1968. *Other developed countries where chemical products were excluded from patentability* till very recently are: USSR, Finland, Sweden, GDR, Brazil, Switzerland, Spain, Poland and Czechoslovakia.

Products excluded from Patentability

a. *Food Products*
Australia, China, Canada, Brazil, Japan, Spain, Switzerland, Korea, GDR, Hungary, Poland, USSR

b. *Pharmaceutical Products*
China, Japan, Spain, Switzerland, GDR, Hungary, Poland, Turkey, Argentina, USSR, Brazil, Pakistan.

c. *Plant varieties or new kinds of animals*
Austria, China, Canada, Brazil, Denmark, Finland, France, Sweden, Norway, UK, Poland, Switzerland, USSR and many other countries.

d. *Nuclear materials, atomic energy/weapons*
Japan, USA, China, Czechoslovakia, GDR, Poland, Brazil, Romania.

Countries which exclude both pharmaceutical (and food) products and processes from patentability include : Australia, Brazil, New Zealand, Thailand, Turkey and Italy (till 1984). UK had process patents only between 1918 and 1949 and introduced product patents only after achieving substantial technological self-reliance.

* *Term of the patent*: Different countries have had varied periods of patent protection for different categories of products which have varied from 5 to 20 years. For example Argentina, China, Turkey, Venezuela and Yugoslavia provided 5-10 years protection for certain products, while for others it was 11-15 years. Thus most developed countries during their industrialization phase had provisions in their patent laws similar to those in the Indian Patents Act, 1970, so why this pressure on us to change!

Inequity between developed and developing countries in regard to Patent Protection

In 1988, out of about 3.5 million live patents, less than 1 percent (about 30,000) were held by citizens of developing countries. And out of these 3.5 million patents, less than 5 percent are likely to be commercialized in the developing countries; the rest are there just to block and prevent others from using this knowledge. In India between 1947 and 1957, 1704 patents were granted for drugs and pharmaceuticals out of which almost 95 percent were held by foreign citizens and not even 1 percent were commercialized in India. Thus for developing countries the patent system is a very unequal relationship, and special provisions have to be made to protect national interest.

The Indian Drugs and Pharmaceutical Industry

In view of the great social importance of the pharmaceutical industry, as it makes available essential drugs and medicines needed for health care, the Indian Patents Act, 1970, has special provisions for this sector, which include:

- Product patents not patentable;
- Term of process patents limited to 7 years from the date of filing or 5 years from the date of sealing of patents.
- License of right and compulsory licensing.

Present status

These provisions have been important factors in the phenomenal growth of the industry in the post 1970 period, which is depicted in Table 1.

Table 1 : Growth of Indian Pharmaceutical Industry

(Rs in million)

		1970	1980
1.	Investment	1500	8000
2.	Production		
	(b) Bulk Drugs	500	5300
3.	Imports	230	3490
4.	Exports	97.5	4000
	(a) Formulations	72.2	1570
	(b) Bulk Drugs	25.3	2430
	(c) As % of indigenous production	4.0	15.0
	(d) Some developed countries to which exported : USSR	USA	F.R.G.
5.	Research & Development		
	(a) Expenditure	40	500
	(b) No. of units approved by the Dept. of Science & Technology	--	77

The achievements of industry may be summed up as follows:

- Practically self-sufficient in formulation production with no import allowed or necessary except for some very recently introduced life-saving drugs.

- Around 65-70 percent of bulk drugs are indigenously made, exports of both of bulk drugs and formulations have increased sharply; imports have also gone up but a sizeable portion of these are export related; export is to both developed and developing countries;
- Self-reliant technology base established and new drugs introduced abroad made available in India in 3-4 years by local production at affordable prices; in fact prices of recent drugs in India are amongst the lowest in the world (Table 2).

Table 2 : Comparative Drug Prices (1988) = (Wholesale price in India. Rs. per pack of 10's)

		India	U.K
Allopurinol	100 mg	5.84	30.3
Atenolol	100 mg	11.29	61.15
Cimetidine	200 mg	6.77	36.40
Captopril	25 mg	15.45	58.56
Diltiazem	60 mg	15.26	40.89
Haloperidol	5 mg	13.58	41.16
Mebendazole	6X100 mg	4.88	37.92
Naproxen	250 mg	12.76	31.07
Nifedipine	10 mg	3.82	29.90
Piroxicam	6X20 mg	7.20	184.75
Ranitidine	150 mg	16.15	121.67

Competitive character of the industry

Due to great pressure for process innovation and modification, the industry has become highly competitive. A number of processes developed by it are novel, but on account of the short term of patent protection allowed in India, patents have not been filed for all of them. Processes for which patents have been filed in India, Europe, and USA include: amitriptyline, catapress, ciprofloxacin, colchicine, doxycycline, indomethacin, norfloxacin, ranitidine.

The competitive character of the industry is also shown by the lower prices of most of the recently introduced drugs which are now manufactured in India as compared to the international prices (Table 2) and short time-lag of 3-4 years between introduction of a drug abroad and its indigenous manufacture.

Self-reliant — technology base

The Indian industry has been able to establish a strong self-reliant process technology base (particularly for synthetic and phyto-therapeutics) and is able to manufacture and market drugs requiring multi-step synthesis/isolation from basic stages by new innovative processes. Indian industry has introduced around 80 new bulk drugs , some of which are very recent discoveries, whose patents are still valid and hence required development of alternative processes. Out of a total of about 500 bulk drugs which Indian industry uses for formulation, over 50% are manufactured indigenously, which is creditable and comparable to the performance of many industrialized countries.

Pressure on India to change the Patents Act of 1970

As Indian industry, particularly pharmaceutical industry, is able to make most of the country's essential requirements and is also becoming a strong competitor in the international market, there is intense pressure on the Indian government to change its patent laws. The demands of the industrialized countries are: (a) introduction of products patents; (b) no patent exclusion, not even for production required for health care; (c) uniform period of patent protection of 20 years; (d) no compulsory licensing or license of right and (e) to have to a dense for reversal of burden of proof. These are the very chances which have helped the Indian industry to grow and it is not difficult to imagine the setback to public health and other societal needs if we go back on these changes. In many references to Indian exports it is implied that Indian industry is infringing international patents -- this is not true as, Indian exporters have to conform to the patent laws of the country where they operate.

Biotechnology and related issues

Biotechnology, broadly defined, includes techniques that use living organisms, or components thereof, to make or modify products, to improve plants or animals or to develop micro-organisms for specific uses. The discovery of new ways of manipulating some of the basic biological processes, which include recombinant DNA techniques, cell-fusion techniques, hybridoma and monoclonal antibody technology, have brought into forefront the vast potential which the ability to manipulate genetic material has, and has greatly heightened the interest in the commercial use of biotechnological processes and living organisms. In view of this, extension of patent protection to biotechnology-based products and processes has become a subject of much interest and debate, in fact of controversy. We have, therefore, to consider our options and position very carefully. Before analyzing these it would be useful to consider some special features of the present biotechnology industry.

1. The wide-ranging potential applications of biotechnology are very relevant to the major needs and problems of developing countries which include health, food and malnutrition, environment, pollution and energy. Biotechnology being highly knowledge -based and skill- intensive but not capital-intensive is particularly suitable for adoption by the 3rd world countries, including India. We have a reasonably well trained scientific manpower in this field and could make a headway on our own. But as it is we are a late entrant in this field. And if the free flow of knowledge or technology and product development are restricted because of Patent-Protection, our advancement will be greatly hampered.
2. Much of gene cloning work is done with and on naturally occurring genes, and thus depends on what exists in nature; in the case of plants, the developing tropical countries are the home of many important germ plasms; this gene pool belongs to the whole of mankind and should be freely available. Any restriction on its uses would go against the well accepted principle of free sharing of this common heritage. Moreover, the basic processes of genetic recombination and cell-fusion have occurred and continue to occur in nature ever since life appeared on this planet, and have in fact been the basis of evolution. So these processes are not novel and as such the proprietary claims to patent them is questionable.
3. Although the industrial revolution in biotechnology started with a broadbase of a large number of small companies, due to mergers and take-overs the industry is getting concentrated in the hands of a few

multinational corporations, and unrestricted patenting and exclusivity can lead to creation of monopoly regimes. And monopoly regimes controlling living forms, human health and plants can have far-reaching economic, social and ethical implications.

4. If a process is available for a wider public use, it is amenable to better moral, social and scientific control to regulate its use than if it is a monopoly.

In view of the what has been said above, we should consider our position, options and interest very carefully and not do anything which will compromise our national interest. The subject of patents has been under intense debate in industrialized countries, and even amongst them there is considerable divergence of views between US and the Japanese/EEC stand on this issue. The EEC recommendation as given in Articles 52 & 53 of the European Patent Convention is:

Patentable Inventions

1. European patents shall be granted for any inventions which are susceptible of industrial application, which are new and which involve an inventive step.
2. Methods for treatment of the human or animal body by surgery or therapy and diagnostic methods practised on the human or animal body shall not be regarded as inventions which are susceptible of industrial application within the meaning of paragraph 1. This provision shall not apply to products, in particular substances or compositions, for use in any of these methods.

Exceptions to patentability

1. Inventions the publication or exploitation of which would be contrary to public order or morality.
2. Plant or animal varieties or essentially biological processes for the production of plants or animals; this provision does not apply to microbiological processes or the products thereof.

The provisions in the Indian Patents Act, 1970, concerning biotechnology and animal and plants are the following:-

Chapter II: Inventions not patentable

3 (h) A method of agriculture or horticulture. 3 (i) Any process for the medicinal, surgical, curative, prophylactic or other treatment of animals or plants to render them free of disease or to increase their economic value or that of their products. 5 (a) Only methods or processes of manufacture patentable for substances intended for use as food, drug or medicine.

Legally interpreted, the Indian Patent Laws are quite similar to the stand of EEC on this issue. There is, however, some ambiguity about the patentability of life forms, and it may be useful to introduce amendment to make the law unambiguous.

Summary

1. A study of the evolution of the world patent system would show that the patent laws of different countries have varied greatly and each country has enacted laws which subserve its national and public interests best.
2. The Indian Patents Act, 1970, with special provisions for foods, drugs, medicines and related chemical

processing industries, has greatly promoted the growth and development of indigenous chemical-based industries and it is in national interest not to change these laws at present.

3. All biotechnology-based products and processes and life forms should be excluded from patentability for the present.
4. An alternative system of rewards and awards should be devised to provide incentives to inventors, such as royalty on the acreage under cultivation for any new plant variety developed or on sales turnover, for any new drug discovered, but the licensing must be non-exclusive.
5. The industrial scene is changing at a very fast rate on account of rapid advances in science and technology, particularly in the areas of biotechnology, electronics and informatics. It is important to continuously monitor and analyse the impact of our patent laws on our industrial development in these fields so that our national/public interest is not in any way compromised. However, any change in the existing patent laws should be considered only after proper open discussion and public debate.

Conclusion

We must not be pushed by international or multinational corporations and must make our own decisions about intellectual property rights and patent laws. Wisdom lies in striking a balance between different conflicting interests, keeping national and public interest supreme.

Technology: Estrangement and Enlargement

D.P. Chattopadhyaya

I

Man is said to be the only tool-using animal. One need not necessarily feel cramped by this "essentialist" definition in one's exploration of the technological aspects or paradigms of man as a tool-using animal. In this paper I would like to show the seemingly two opposite faces of technological culture — estrangement and enlargement.

II

Tools are ordinarily deemed to be external to man, his embodied existence. But this view has been questioned on subtle philosophical grounds. For example, it has been rightly pointed out that it is difficult to draw a line of demarcation between where my body ends and where the world begins. The question may be formulated from the reverse end. Where the world ends and where the embodied existence of man or of my own self begins? The stimuli which irritate human body are not necessarily external to it. What is external to it can hardly be indicated without determining conditions and with reasonable correctness. What is internal to it? Sometimes we are stimulated or irritated by our own idea, imagination or recollection. At times clearly identifiable external stimuli or irritations fail to affect our body. Both sensitivity and sensibility of human beings are apparently not uniform and therefore cannot be satisfactorily explained in terms of physical causation as ordinarily understood.

Here I do not propose to go into the details of these philosophical subtleties. Rather I propose to take a commonsense point of view.

We use various types of tools, hand tools, machine tools, and other sophisticated instruments. It is well known that hand tools in different ages, pre-historical and historical, had not been of the same type. The materials out of which they used to be made are found to be of different kinds — stone, bone, metal, etc. Similarly, machine tools are of different levels of precision and sophistication. Neither precision nor sophistication knows any limit. In brief, historically or cross-culturally speaking, improvability of tools is boundless. Often we speak of hand tools. But hardly we hear of hand as tool. What makes a "thing" tool? A piece of stone is regarded as a knife if it can be used for the purpose of cutting certain other things. All knives need not be very sharp or sharp in order to be recognised as knives. A blunt knife is also a knife. What is more, a blunt knife, whether of stone or of metal, can be easily used for the purposes of pressing, breaking, cutting, piecing, etc. There are tools which are multipurpose.[1]

What about our hands as tools? In a sense my hands are parts of my body. But these are not as essential as heart and kidney, for example. Without the latter one cannot live. But many human beings are there without hands. In this specific sense our hands are not purely internal or essential to our being. What is more important to note in this connection is that our hands can be used for many purposes which are ordinarily performed by different tools. By hands we can press, break; cut, push, pull, throw, sew, etc. In a sense hands, being parts as these are of our body, enable us to do or achieve many things. Hands take our body out in the world to bring about some changes in it. There is an ambivalence in manual works. Compared to hand tools, hands as tools, are closer to our psycho-somatic system and therefore can be handled more easily, quickly and in a refined way. At the same time, being parts of our body, hands cannot acquire that objective and transferable characters which increase the use-value of (non-manual) tools. This shows what I call the ambivalent character of hands as tools. Undoubtedly, we own our hands or, to put the matter differently, hands belong to us. Hands are inseparable parts/organs of a normal human body. Loss of hand, organic or functional, is not like dysfunctional kidney or ruptured heart. Given the latter deformity, one cannot live. But this is not the case with the loss of hand. Hands are potent tools and help us in enlarging our world. Without hands we are handicapped. But without hands we are not dead and do not cease to be what we basically are.

One important peculiarity of hand as tool is its distinction from other forms of tactile sense. Besides hands we have various other parts of body which are equipped to give us sensations of touch and sensibilia thereof. Hands not only have tactile qualities in them but also have the additional and, in a sense, peculiar capacity of performing certain things. It is true that some of the activities of hands like writing, drawing and painting, pulling, pushing etc., can be transferred to and performed by feet through sustained exercise. Even then it may be pointed out that strength, delicacy and variety of manual works cannot be transferred to feet in toto.

Perhaps a more general point may be made in this connection. No part of the body is equipped to duplicate exactly the mode of reception and function of any other part of it.

Human body is often said to be a *tertium quid* between our self-existence and the world. The statement, though somewhat metaphorical in character, highlights one very important aspect of human existence. Body situates us in the world. Our world is enlarged by our body. What is more body brings the world in a screened form to us. Bodily screening of the world yields different qualities of it, tactile, auditory, visual, etc. Simultaneously, while my body situates me in the world I am partly thrown into the world. In and through my work my being-in-the-world may be primary. That mode of my self-existence dominantly is a mode of estrangement. I must hasten to add here that my mode of self-estrangement like that of self-enlargement cannot be purely one-sided. Even the consciousness of one-sidedness or primacy of this mode or that mode is nested in self-existence itself.

III

Can eyes, like hands, be tools (in their relation to body)? The answer to this question may be constructed both ways, Yes and No. Eyes are said to be most informative, if not influential as well, sense organ. Their reach is far. The qualities of world they can capture are varied. What the naked eyes cannot see the tutored or trained ones can. Natural capacities of eyes can be augmented in two different but related ways, by training and by aids and adjuncts. The primitive astronomer, for example, trained by his preceptor or father could see the stars in a more meaningful way than the rest. Everybody, except the blind, can observe. But evidently all persons of normal vision do not and cannot see the world around and above them in the same way. Whether one observes stars or flora or fauna as a layman or as a professional makes all the difference in what one observes. It is true that observation is not as regulated as experimentation. It is relatively free. But it is not aimlessly free. In a sense the aim of observation set a limit to what the observer would see. Interest and attention of the observer converge on his aim and regulate the range of the objects of observation.

Whatever we see, big or small, takes us, into the world. Vision may be aided or unaided. Macroscopic or telescopic vision takes us to the far and wide parts of the universe and enables us to see seemingly small celestial bodies in much larger forms. And on the power of the telescope used depends the (visible) forms or the largeness of the distant bodies. Similarly, in microphysics and microbiology, for example, we can seek very small particles and genes in considerably enlarged forms. With the passage of every generation our microscopes and telescopes are becoming increasingly powerful. Enabling us to bring the distant near to us and to present the smaller things, including the invisible ones, in much enlarged forms. In a way aided vision is a manipulated vision. What normal eyes cannot see are made available to them by external aids.

Eye in a sense is not an external aid. It is an internal organ, enclosed in a socket. But it not as internal as our pancreas or lung, for example, is. Even our "internal" eyes take us to the world outside us. It provides us an easy passage to external world. When it is trained and, moreover, aided by instruments like telescope and microscope, we are ushered into an enlarged world.

To understand the estranging effect of visual perception it is useful to think of the mind of the blind and contrast it from that of the person with vision. It is said of the people who are deaf, dumb and blind that, deprived of the inflow of information through the said organs, they are largely compensated by the psycho-somatic nature of human body. Many persons without one or two or even more sense organs are found to have achieved extraordinary success in different fields of creativity — artistic, musical and scientific. The classic case of Hellen Keller is often referred to in this connection to highlight the point that out of meagre (information) input massive output is possible. The interesting point to note here is that this is so only in the case of human beings. There is something peculiar in human-mind or self which accounts for this extraordinary or disproportionate relation between input and output in the human-system.

Close observation of almost every person without vision shows some or other remarkable gift in him or her. Loss of vision seems to account for augmentation and refinement of interest in and attention to the information received from without. Besides, it tends to make one's mind more and more imaginative and reflective. Visual perception, more than any other form of perception, entails a sort of dissipation of mental energy. Since of all forms of perception, the visual one plays a very important role in our mental life, the amount of energy it dissipates is considerable. When this dissipation is arrested, the eye of the mind turns inward. Primacy of inlook, loss or non-availability of outlook, makes our "internal survey" extensive, intensive and protensive. The results of this type of "survey" brings out a fundamental change in our internal life and imparts extraordinary excellence, delicacy and sensitivity to it.[2]

By tracing another line of consideration the technological aspects, enlargement as well as estrangement, of visual perception may be clarified. When a poor man cools his body by moving his handfan, made of palm leaf, for example, he can see for himself what he is doing and what is its result. Similarly, when he sits by burning leaves or logs of wood in winter to warm up his body, he knows what he is doing and why. In this sort of cases cooling and warming are nothing mysterious to the poor Indian villager. But suppose we take the same villager to a five-star hotel in some big town, Delhi or Bombay, where rooms are centrally airconditioned, heated and cooled. In this case the man, uninitiated in the technology of central airconditioning, though feels the pleasure of being cooled or warmed, does not know exactly how this welcome effect is being brought about "for him". The invisibility or remoteness of the technological operation leading to his perception makes him feel somewhat puzzled or bewildered. Had he been accustomed long to it, i.e., if he would have known the work of the concerned technology, his sense of puzzlement or bewilderment would have been considerably less. The difference between manual technology and invisible technology is bound to be there. But technology in this case is faceless. Its impact is bound to appear somewhat bizarre. The sense of certainty and security attached to visual perception of proximate objects, when missing, make us feel uneasy even in our pleasure and happiness.

IV

The last point brings to our notice, although in a somewhat round-about way, the result of psychological gap between our experience and choice. Rightly understood, this gap is rooted in technology.

If one's experience, pleasant or painful, has nothing to do with the concerned one's choice, one is bound to feel more or less helpless. For example, a villager, used to long walk for moving from one place to another, and hard work for earning his square meal is obliged to live in an extremely modern urban area, he cannot avoid a sense of cultural estrangement, if not shock. Fast traffic, large crowd, loud and varied noise, and unknown map of the place make it difficult for him to walk down to his destination, distant and not-so-distant. Whether he likes it or not, he is advised to go by taxi, bus, underground train or some other easily available transport. Secondly, when a person used to go to market to buy his provisions, to cook his own food and eat at home, is forced by the circumstances of urban living, to eateries for fast food, he feels more or less derailed, despite the ease he experiences in his new set up of life. Ordinarily speaking, he is expected to be happy with the things made easily available to him. But as a matter of act, his experience, phenomenologically speaking, is otherwise.[3]

In a primitive or relatively rural form of life, people are found to be directly related to most of the things they produce and consume. In a way Rousseau recommends this form of life as ideal. In a slightly different way Gandhi's self-reliant society rests on a comparable sort of assumption. In their ideal forms of life not only production and consumption but also choice and experience are intended to be brought closer to each other. However, given the constraints of the prevailing stage of technology, it is difficult, but not quite impossible, to narrow down the said gaps.

This point gives rise to another interesting collateral issue. Does a particular former stage of technology lead to a particular form of life and, by implication, rule out other forms? Technology is said to be *neutral* between different or alternative forms of life. But is it really so? Is one justified in espousing a deterministic view of technology? Do we willingly surrender to this suggestions of technology or succumb to its dictation? In brief, the claim of neutral technology and the thesis of technological determinism do not seem to go well together.

That placed or situated in the same technological society we, different individuals, can, in fact do, live differently goes without saying. For example, in a modern society one can live in an outdated way. A man of Gandhi's character can, as he did, go to Buckingham Palace like a "half-naked fakir" and with his favourite pet goat, Nirmala. The same sort of (strictly historical) stories is current about Ishwar Chandra Vidyasagar and his British bosses of the 19th century Calcutta. This shows that there are few among us who can carry their own chosen forms of life to anywhere they go irrespective of the latter's economic and technological characters.[4]

But these exceptional cases perhaps do not negate the basic thesis of technological determinism. How our houses are built, how we cook our foods. Foodstuff themselves, cooking fuel, etc. cannot be entirely determined by us. Many of these items are more or less determined by the time, type of culture, type of dominant technology, etc., amidst which we are obliged to live. The place of our dwelling itself, town or village, is not exclusively a matter of choice for us. Given a particular income-bracket and family obligations, one may have to move from village to town or town to village or from one part of town to another. In a fast changing society one has little option about one's place and type of work. Technology and market are great (not necessarily noble) equalisers. The equalising forces of market and technology that I am alluding to are also responsible for the gap between our choice and experience, action and result.

The concept of *neutral* technology is ideal-typical. It suggests a highly idealised situation and the role of technology therein. To change the analogy, it is like a frictionless machine. Strictly speaking, machine by its nature cannot be absolutely friction-free. It can be so only relatively.

If one looks into the specifics of the functions or the modes of utilisation of a particular technology, it becomes clear that it does not work in a vacuum, political, economic or social. No-ownership theory of technology is untenable. It is contrary to the facts of political and economic life. Technology, while patented, is owned by this or that legal person, individual, institution or government. It is true that owner may let others to possess and use it on payment of the stipulated fees. Technology, owned or possessed, is put to some purpose or other which, on scrutiny, is found to be more or less *partisan* in character. The technology, controlled by a person or a group, is not ordinarily allowed to function against the interest of the owner. Ordinarily speaking, my hands and mouth, manual works, writing and utterances, are intended to express and further my own will. But at times, we come across situations produced by our works and words which go against our own interest. Not infrequently we hear some such expressions as "every time he opens his mouth, he creates problem for himself".

In larger social and economic context in which many individuals and groups act and interact there is bound to be some effects not specifically intended by anybody. "Frankenstein" is not necessarily an imaginary character. Quite a number of machines or, at any rate, some functions of many machines prove inconsistent with the interests of their masters. The general point may be illustrated in terms of the relation between industrialisation and pollution. It would be unfair to assert that every industrialist is a wilful polluter of the environment. This sort of assertion is traceable to the conspiracy theory in economic and sociology. It is fallacious to maintain that economic development is destined to subvert the balance of the eco-system. At the same time, the scrutiny of the anatomy of the economics of industrial development shows that it contains, knowingly or unknowingly, the seeds within itself which lead to deforestation, pollution of air and water, or some such harmful effects on human.

The ambivalent or bivalent, beneficial or malficial, characters of technology are clearly in conflict with the neutrality claim preferred on its behalf. For example, the publicity campaign, launched and financed by a government or a company, is undoubtedly aimed to promote the acceptability of their goods, services and policies among the concerned people. A responsible government or company can hardly afford to spend on a project which is contrary to its interest. If the chambers of commerce or the producers' cartels try to maximise their profit within a given time-frame, the consumers' groups also try to protect, legally, administratively and otherwise, what they consider their own legitimate interests.

But in a market situation neither the producers nor the consumers can actually behave in a very homogenous manner. In a way every producer, in spite of his affiliation to or association with a group, tries to carve out a specially profitable position for himself. On the other hand, every consumer tries to be most faithful to his own preference-schedule, despite the fact that his preferences are not quite independent of those of others. The much talked of sovereignty of the consumer is more notional than actual.[5]

V

Utility of technology is transparent and manifold. From hand as tool to hand tool and machine tool, every technique saves our labour or improves the quality of the product of labour or both. Different forms of biotechnology enabling the experts to undertake hitherto unimaginable forms of surgery like heart transplantation and kidney transplantation have considerably reduced human suffering and improved the quality of life and living. The utility of technology must not make us blind to what the radical forms of technology and their proliferation are likely to create for us. Increasing dependence on technology, while saving our body from harmful effects of hard labour and relieving our mind from avoidable worry and anxiety, silently and steadily, takes away our enthusiasm, initiative and larger sympathies for life. Excessive exposure to technology results in under-utilisation of our immediate psycho-somatic capacities. Increasing non-use or sub-normal use of the limbs of body and other creative propensities of mind, conginitive, emotive and volitive, tend to kill our initiatve and make our living uninteresting. This technology, originally intended to save our labour and enlarge our life, in fact takes us, at least the rich among us, to the land of lotus eaters. In a way it tends to make us relatively indolent.

What is sociologically even worse is this. Technological society, if not otherwise corrected, also tends to make our common life fragmentary. It is true that different forms of communication and electronic technology can brings us closer and, what is more, try to put us together. But, at the same time, it is to be noted that this "technological intimacy" in effect proves more illusory than real. Increasing use of technology makes human relations faceless. Face-to-face relation gradually becomes a dispensable item, if not wilfully avoided experience, in the highly developed technological society. For example, in the name of privacy, we, who live in a same apartment building try to avoid visiting each other. For enjoying the supposed privileges of privacy and anonymity we develop the habit of living apart. Our existence becomes thus fragmented in the same building. In the name of enjoying freedom our life gets confined within the walls of our own apartment. In quest of freedom we are shut up in our small household.

Another and more disturbing feature of technological society is erosion of democratic spirit. Without enthusiasm, interest, sympathy and initiative on the parts of individual citizens, the democracy becomes an unlived socio-political shell. If technology is allowed to separate us from one another, making us indifferent to others' weals and woes, we, in our turn, cannot reasonably expect to be comforted by others in our moments of distress, despair and suffering. Technology may not be directly responsible for all these ills of technological society, but its stance of neutrality being untenable, we cannot totally delink our negative experiences from the technological infrastructure of our contemporary society.[6]

Another negative feature of technological society is to be found in its forms of sharp social conflict. While, on the one hand, the richer sections can live the life of a lotus eater, roll in affluent indolence and can afford to be indifferent to others' sorrows, the poorer sections, on the other hand, are brought together by the forces of circumstances and due to synmpathies born of common suffering. The great divide between the rich lotus eaters and the poor fire eaters might have been dramatised by the radical socialist. But the core of its truth can hardly be dismissed in a lighter vein. It is true that the Marxist theses of gradual polarisation and increasing pauperisation have not come true in the way Marx and his followers anticipated in the late last century and the early this century, but another collateral division between the poor and the rich has come to stay.[7]

By highlighting the creative impulse of the market economy its competitive cruelty can hardly be ignored. Technology is undoubtedly the most close ally of market. But there is a view of market which tries to portray it as an institutional symbol of benign nature. Like nature, it has been argued, market promotes and sustains the forces of natural justice. By implication, this line of argument tries to depict technology, the trusted ally of market, also as naturally good and just. We are given to understand that technology is an artificial limb purported to extend the functional reach of human limbs. Technology is basically understood in the analogy of natural human organs, sensory and motor. This view of technology and market, resting as it does on the concerned theoretician's sympathy, fails to take due notice of the cruel competitive character of the invisible bosses of market and equally invisible masters of technology. It is difficult to believe on the basis of available evidence that their "invisible hands" are promotive of co-operation between the poor and the rich, between those who stand behind the supply side and those behind the demand side.

Studies in the bivalent characters of or the dialectical interaction between utility and disutility, between attraction and repulsion, of technology reveal two very important things. First, in its initial bid to help man, to enlarge the area of freedom of his life, technology gradually starts harming him, estranging him from himself. Starting work as extended limbs of man, gradually it takes the character of the self-scratching claws of the big and wild cat, inflicting unwittingly festering wounds on its own body. This analogy is purported to convey the correct impression that the development of technology in its relation to human life and society displays a persistent tendency to overtake the latter and supervene over it. Strictly speaking, I concede, this dominance of technology over man is not inherent in the process of technological development. I do not propose to give a devilish image of technology. Technology "in-itself" is *not* a devil. Technology is what we make it to be in our ambivalent relation to it, simaltaneously or alternatively getting close to it and withdrawing from it, using it and highlighting the importance of its non-use.

Secondly, technology as tool *of* man and *for* man, in the process of its development, starts acquiring some characteristics and a momentum of its own which are bound to give one the impression that technology, beyond a point, proves *autonomous*. This autonomy may be construed in two different ways, positively as well as negatively. The "Frankenstein" image of technology is patently negative. It is the revolt of the machine made by man against man himself. I for one do not like to press this negative image of technology as a moral criticism of technology. I see it primarily as a sociological phenomenon of the modern time, especially of the modern man who is strongly drawn to the charm and comforts, partly real and partly illusory, of technology. At the same time, one feels inclined to recognise an element of inevitabilism in the process of technological development. Once we opt for the best possible fruits of high technology, we are fated to reap the attending bitter harvest. There is no easy escape route between "the devil" and "the divine".

Neither the idyllic picture of the pre-technological society nor the devilish vulgarisation of it should be taken in a literal sense. The "autonomous and runaway" character of technology is certainly controllable to a significant extent. This controllability depends much upon our internal strength and resources. We must manage and control our needs and try to arrest their continuous multiplication. Uncontrolled needs tend to degenerate into greeds. Our external resources ordinarily do not prove adequate to satisfy these imitative and runaway greeds. It is like the typical consumer's dilemma. Influenced by publicity and fellow consumers, when the consumer tries to follow others' consumption patterns, he may quickly exhaust his own resources and become a bankrupt. Alternatively, he may control his high-consumption propensity and in the process incur the wrath of family members and the disdain of the neighbours and the friends.[8]

But this sort of dilemmas one comes across not only in the context of one's relation to technology but also in various other social contexts of life. The main attraction of the Gandhian way of life lies in its emphasis on the *self-reliant* nature of life, both individual and social. If the individual does not succumb to the temptation of having all those commodities and comforts which his neighbours and friends have and tries to manage his affairs to the best of his own ability, without borrowing from others and depending very much on others, he may be able to develop over time a distinct life-style of his own. It may be austere but not poverty-stricken. However, this is not to deny that our individual life is not impervious to our social existence. In a hi-tech society one cannot afford to have the imaginary charms of the pre-technological idyllic state of nature. The ascetic's option may have some moral significance in his own limited context but to draw a general moral lesson from it for the society as a whole will be a misplaced and avoidable foolishness.

The Gandhian solution of the dilemma posed by the process of hi-tech development is very much relevant to the large sphere of society, beyond the life of the individual. If we are not to give up ourselves to the supposed type of inevitablism, we have to stand firm on our own feet and refuse to be swept away by the flying fashion of the time and the strong winds of imitative consumerism. This is not an impossible proposition. To realise its truth one must be prepared to fall back upon one's own internal resources. Though it is true that the supply lines of internal resources run beyond the limits of our individual life, yet we would be well advised to remember that the limits between the two spheres, individual and social, are changeable and enlargeable. Therefore, one need not feel unduly cramped by one's imaginary sense of helplessness. Given determination and self-control, we can overcome, to start with, many of the constraints and limits within which we find ourselves situated. Technology is not the devil to be afraid of. We, as its architect, can tame and control it to a substantial extent. Like all sorts of controls, every act of controlling *other* must be preceded by the disciplined acts of *self*-control.[9]

The question may be raised whether the Gandhian ideal of self-controlled consumerism is viable in a hi-tech market situation. The autonomy that is ascribed to the hi-tech development process seems to be represent in fuller measure in market itself. Given that character of market, are we free to follow our idealistic path, defined in terms of self-control, austerity and self-reliance? Any plausible answer to this question appears to rest on the view of market acceptable to us.

Broadly speaking, I am alluding two different views of market. Market may be understood as an extension of nature itself. Like nature, it is said to have its own laws, the most important ones being those of demand and supply. Causes and effects of nature are generally found to be equipotent unless in between them there intervenes some or other additive or subtractive force or phenomenon. Analogously, it is believed that the economic causation evident in market is basically illustrative of the balanced and balancing characteristics of nature. The self-determining image of nature, marked by its self-contained laws, need not be necessarily traced only to the Newtonian paradigm of nature which, historically speaking, have influenced the thought of many European thinkers like Locke, Hume, Harvey, Adam Smith and Malthus in their respective fields of study, philosophy, political economy and physiology. Mechanical or pro-naturalist paradigm is found in other cultures and in other schools of thought as well. For example, the non-teleological, *Samkhya* paradigm, if developed and applied to market analysis, it is bound to take a distinctly pro-naturalist form.

But one need not necessarily be pro-naturalist in one's construal of the concept of market. When we speak of "free market" the adjective "free" need not be taken in an naturalist sense, i.e., in terms of mere absence of external constraint. market may be free because it is a field of free enterprises and activities of different individuals. It may be readily conceded that individuals, exposed to societal facts and forces, exhibit common traits of their minds and, what is more, their actions and interactions in the context of market are subject to the laws of market. But this concession does not entail denial of free disposition, initiative and action on the part of individuals who at times co-operate and at times are engaged in conflict between themselves. Conflicting and co-operative activities need not be taken as primitive or absolutely autonomous. They depend much upon individuals' own perception of their situation, comprising their societal facts and forces. It is to be further added here that societal facts and forces are not self-propelling independent entities. Their formation and transformation, in turn, are open to the influence of dispositions and actions of the concerned individuals. In brief, methodological individualism and methodological socialism are mutually supportive of one another. Depending on the nature of the objects of our analysis we give primacy to one approach at one time and another approach at a different time. Methodological pluralism does not mean *ad holism.* It only means the researcher's readiness to be fair to the facts under study.

The anti-naturalist idea of nature and market highlights the place and importance of imponderables in both. How market or nature would behave, under specific circumstances, cannot precisely be anticipated beforehand. This limitation of our understanding is not purely due to our nature. It has to do something with the structure of nature itself and that of market as well. It is said that neither nature nor market has been fixedly structured by some eternal and immutable laws. On the contrary, we find from the history of natural knowledge and human economic activities that certain things not found in earlier times have come to light or to human notice much later. To attribute all these new or emerging characteristics to what is called the historical *a priori* would be to deny altogether or at least minimise the importance of free human activities and results thereof.[10]

Besides, the pro-naturalist view of nature is obliged to endow nature with some counter-intuitive properties. Can nature be deemed to be naturally rational? Is rationality native to the very structure of nature? Is this rationality directly lent by nature to market? If so, how? Are we then to believe that human nature, *a la* physical nature, is causally determined, despite its contrary surface look? The defender of the pro-naturalist concept of nature also invites us to believe that nature is not only rationality-ensuring but also justice-dispensing. In other words, we are asked to believe that full play of market forces almost automatically give rise to not only rational action in economic life but also to just action in moral life. The premises which are suppressed here, if spelt out, read like this. Exposed to natural influences, human beings have in them certain instincts and inclinations of actions which, duly followed and acted upon, augment our pleasure, minimise our pain, and show the right ends as well as means to be espoused. Further, the natural history of mankind is said to be replete with illustrations in the forms of traditions and customs which, if consistently followed, assure our freedom from errors of judgement and mistakes in action. The pro-naturalist thus tries to prove nature as the infallible master of mankind in every sphere of life, political, economic and technological.

If technology, like economy, is believed to be under the constant and unfailing influence of secret natural causation, one need not be unduly afraid of "Frankenstein" technology. Nor are we in that case required to be extra-careful in planning the right place and course for technology in relation to human welfare. In brief, we are told not be worried about technology because it, like nature, is destined to take care of itself.

Gandhi, with his strong pantheistic leaning and religious feeling, discards the pro-naturalist account of market. Also he is opposed to a collectivistic or even welfaristic conception of market. A fierce defender of the value of individual freedom, he wants that man should mainly depend on his inner working habits, moral impulses, and spiritual resources. One must not think that he endorses the Protestant view of the managerial craze for work and extra-work is not what man should accept as his ideal. This work-culture is calculated to make man more and more acquisitive, not really self-reliant. Gandhi believes that man can exercise his freedom to work in such a way that he will be able to retain effectively his control over the means or technology of productive work and will not be devoured or determined by the latter. This libertarian view of work substantially hinges on a spiritual view of nature and on the belief that man is not necessarily fated to be shaped by the forces of market.

Gandhi's views on work and market are substantially inconsistent with the autonomy claim of technology. The autonomous technology not only interferes with the ideal of individual freedom and dignity of labour but also cuts into the very possibility of internal development of man. If technology can develop and can go on indefinitely developing on its own momentum, the realm of human freedom is bound to shrink under its uncontrollable impact. The supposed regardlessness of powerful technology, if comes really true, how man can be self-reliant? Then how man can remain free to form small communes, — *panchayat,* Workers' Co-operative, Producers' Association, Forum of Trustees? Gandhi is opposed to every form of social, political and economic collectivism. Whenever collectivity is formed in a large way, it is motivated by a strong formative impulses and which, in most cases, gives rise to use of technology in a big way. His commitment to the ideal of "small is beautiful" is seriously meant. Otherwise, he fears, perhaps rightly, that big Producers' Guild and equally strong Consumer's Resistance Group are willy-nilly driven to a sort of aggressive collectivism. Even for the sake of a supposed good cause promotion of collectivism is rejected, because as means it is unsupportable. Besides, once it is allowed, its pernicious effects cannot be stopped.

The dominant trend of anarchism evident in Gandhi's thought also proves inconsistent with autonomous and high-tech view of culture. If individuals are really intended to be maximally free, their lives must be as self-reliant as possible. The concerned State should be minimal in its scope and activities and the role of technology subservient to their needs. Gandhi is perceptive enough to point out that the idea of maximal State cannot but summon the services of high-tech for its survival. The low profile of technology is not consistent with the concept of Statism. Positively speaking, the maximal State arrogates to itself so many activities that without the mechanical help of high technology it cannot perform the same in the intended way. The tragedy of the high-tech culture is that once it is allowed to come to the aid of the State needs, it gradually carves out a role for itself almost regardless of the dictates of the State and the needs of the citizens. At least that is Gandhi's apprehension.

That is why in his concept of the ideal State, *Ram Rajya* individuals are expected to be as self-reliant as possible and extremely co-operative between themselves. By implication, Gandhi favours the idea of a co-operative state wherein the role of the state itself would be minimal and most of its activities would be "delegated" to the Farmers' Co-operatives, Traders' Guild, *Panchayat*, etc. To prevent the rise of the cult of the State and also high-tech culture therewith he expects the citizens to take upon themselves the task of controlling their runaway needs and greeds. Without transforming human nature, it is argued, we cannot free ourselves from the dominance of high technology.

VII

The role of institution, like that of technology, has an important bearing on the question of enlargement of human freedom and human estrangement. The state is a legal person and theoretical entity. Its aims are executed by the arms of government and its corresponding activities undertaken and sought to be performed by the concerned institutions, legislative, executive and judicial. The powers of this governmental agencies or institutions derive their legitimacy from the legal or constitutional status of the State. The more powers the executive is given by the State and the less accountable the latter becomes to the legislative and/or the judiciary, the more threatened becomes the freedom of the common citizens in the concerned State. As a mater of fact it is generally found that the relation between institutional power or organisational strength, on the one hand, and human freedom, on the other, turns out to inverse. It is not destined to be so. But when the rulers of the State keep for themselves excessive power and exercise the same through highly institutionalised or organised agencies, the concerned citizens' area of actual freedom starts getting increasingly shrunk.

This assertion perhaps need some qualification. All sorts of power, born of scientific, artistic and spiritual excellence, for example, are not ordinarily antithetical in relation to freedom. Only when these powers are institutionalised in the forms of unquestionable scripture, infallible Papacy or priesthood and State-controlled Academy, the freedom of ideas and battle of ideas become threatened. The premium necessary to insure freedom is not de-institutionalisation of the State powers but to find out ways and means to regulate, right and good persons to man those institutions so that these instruments do not go out of their bounds and turn out to be instruments of tyranny.

A comparable tendency is often found in the life and works of technology. Technology as such is not out there either to enlarge human freedom or to estrange man from himself. The very expression "technology as such" is somewhat misleading and abstract. For it has no life of its own. Its stance of neutrality is more transcendental than functional. As available to us, ordinary mortals, especially the rich and the powerful, its seductive power is too strong to be easily spurned. Since technology cannot be said to have a life of its own, its works and effects cannot be attributed to its own nature. Its nature, good or bad, right or wrong, whatever that may be, is due to its human use and misuse. After all it is brought about in existence and put to work primarily by human beings. If it starts acquiring more powers than it was originally invested with, the fault or responsibility does not primarily lie with it. It is because of the changing nature of our needs, infirmities of our character, and lack of self-control that we allow ourselves to be dominated by technology. It does not dominate us on the basis of its own strength or superiority. In an abstract sense technology is really neutral. If it is found to be vested with a self-perpetuating power within it, i.e, if it proves really autonomous, the responsibility is primarily ours, especially of the masters of institutions and of the state of technology, at a particular period of time and within given culture.

Dominance of technology, like that of institution, on our life is not a one-sided matter. It brings about a sort of schism in our soul. Our personality starts getting split within. In good faith one cannot totally give oneself up to the hands of technology and surrender to its strength and temptations. There seems to be a moral core in every human being which cannot be easily bartered away to and subjugated by the forces inimical to native human freedom.

Underlying this assertion of mine there is an obvious anthropological assumption. In brief, the assumption is, man is basically free and morally inclined. To start with, this assumption may appear philosophical or even metaphysical. How we characterise it does not make much difference to its mode of availability to ourselves. For example, if it is said that free and moral nature of man is an *essentialist* thesis and that it *existentially* found to be aggressive, exploitive and without much love or sympathy for others, I would readily agree because the human history handed down to us is not without war, bloodshed, treachery and other evidence of evil in life, individual and collective. But is it now equally true that history is also abound with the inspiring examples of self-sacrifice, patriotism, love for others and renunciation? Strictly speaking, the panorama of history presents a mixed view. So is our experience of life. Neither easily lends itself to one-sided interpretation.

Perhaps, a general truth can be affirmed without fear of much contradiction. And that is both recorded and unrecorded history is a narrative, though interrupted and even reversed at times by contrary forces, of enlargement of human freedom at the initiative of human beings themselves. In the process they have created for the purpose institutions, technologies etc.

The ambiguity of our very being is reflected in our social existence. It is a well-known philosophical controversy whether our self is basically psychic or somatic, spiritual or embodied. If it is said, as the pan-psychist does so, the whole of our being is psychical or spiritual, we feel like asking "what about our limbs and sense organs?" Undoubtedly I own my body or, negatively speaking, cannot disown my body. But that does not mean I am free to assimilate or internalise whatever I have as my body and limbs. I have already tried to show earlier that our limbs or sense-organs can be made more efficacious with the aids of some tools and machines. That "enlargement" of body, "artificial" or "mechanical", and augmentation of its capacity are not integral parts of my being. In brief, neither body nor technology is fully disclosive of my being. On the contrary, one may rightly point out that we cannot fully express ourselves because of the limitations of our body and its senses. Rightly understood, our body in relation to the world is simultaneously expressive and suppressive, disclosive and enclosive.

The above ambiguity of human existence is replicated, with some disanalogies, in our social existence. Society and its dynamic perspective, history, are certainly expressive of our yearnings for freedom. But that is not the sole identity of what is history or society. Therefore, we speak of both progressive and regressive shifts in history. Therefore, we speak of freedom-promotive and freedom-preventive institutions in society. And that is why we ween a rational arrangement of the institutions in a way so that our freedom is not either unduly curbed or it is allowed to degenerate into licence of anarchy. Even for the preservation and enlargement of freedom we need some institutions.

Rightly understood, there is no incompatibility between institutionalism and individualism, methodological or ideological. Both for the understanding of the meaning of individual action we are obliged to recognise the societal or the institutional facts. On the other hand, institutions and societal facts are theoretical abstractions *sans* their "internal relations" with human dispositions and actions. In brief, individuals cannot be assimilated, without residue, within any institutional framework. Nor it is possible to reduce, without remainder, institutions into individuals, their dispositions and actions.

Analogous ambiguity is evident in the field of man-technology relationship. Man-technology relationship is not face-to-face or direct. It is mediated by societal forms. The forms of mediation are subject to historical change.

It is only in small groups or communities that man's relation with technology, though indirect, is not very remote. Sense of personal control over technology gives man a sort of personal satisfaction. This is due to the less mediated character of relation between man and technology. As I have tried to argue earlier, when relation between the two is extensively mediated, i.e., when man cannot clearly understand how he is being environed or influenced by technology, he feels askance about it and an element of anxiety is found to be there in his relation with technology. When the most used forms of technology of man are found within his primary reference group and he is not overwhelmed by the same, technology proves an ally to him and not alien to his consciousness.[11]

In order to be in peace with himself man wants those forms of technology which are primarily self-preserving in character. When I speak of the "self-preserving" character of technology I am highlighting man's social existence in the synchronised perspective and distinguishing it from the self-exceeding diachronic perspective. Man's social existence, broadly speaking, exhibits two main aspects, self-preserving and self-exceeding. In unsteady balance or tension between the two, man and technology, has direct bearing on the nature and scope of human freedom from the ethical point of view.

VIII

When our technology in the Gandhian sense is modest and controllable, we can easily enter into harmless, if not beneficial, relationship with it. When technology does not harm our relationship with nature and our social environment, it is welcome, for it allows our normal form of life to linger on and does not throw us into vortices of fast life. In other words, modest and small-scale technology minimally affects our normal and healthy relation with nature and other human beings. Cultural life is not unduly disturbed by it. In other words, low-profile man-technology relationship encourages our self-preserving social existence.

Its collateral ethics is deontological. Certain norms or rules are to be followed in order to preserve man-nature and man-technology relationship. If we, armed with aggressive technology, approach nature and start exploiting it, nature is bound to retaliate. Impoverished nature may well famish us both physically and, given time, culturally. So in our own interest we are advised to be in harmonious relation with our natural environment. We must not submit to the contrary prompting of our own nature, its growing needs, and those of technology used by us. This is possible only when use of technology is under effective control of genuinely espoused rules of ethics. It is easy to speak of the advisability of deontological ethics in the context of self-preserving social existence. But to follow it in practice is not so easy.

Here lies the tension between the static ethics addressed to the needs of self-preservation and the dynamic ethics strongly drawn by the changing and growing needs of human nature.

Rightly analysed, the rules of ethics are found to be human, not divine, in their origin. The framers of the rules, being themselves historical and fallible creatures, cannot author any ideals or norms which can be eternal in their efficacy. After all, meant for human needs and subject to the limitations of human authorship, ethical rules cannot acquire for themselves a quite autonomous character. They cannot and do not remain operative for ever. Relevant to and shaped by particular forms of social institutions, they cannot survive for ever. Static ethics of self-preserving existence and evolution of human institutions and morals hardly go together for a long period of time.

Man's existence, however calm it may look at its surface, is always dynamic and potent in its in-goings. While its surface features are relatively steady, its inner drive proves self-exceeding. This may be ascribed to innate freedom of man. The locus of man's freedom being his body and social institutions, he has always to try to see it that he does not somehow get stuck up in the unexercised frame of his body and the unlived shell of his environing institutions. It is only through exercise, inner and outer, that human body remains alive and functional. Institutions, strictly speaking, have no life of their own. Their life consists in human involvement in them and in the resulting propulsion. When human participation in institution is at a low ebb, the inner life of institutions tends to get dried up, resulting in the relative emptiness of the concerned institutions. To stave off this eventuality, degeneration of social life, we are somehow *responsible*. Here lies the necessity and origin of *purposive ethics*.[12]

Unlike deontological ethics, purposive ethics is basically dynamic in its orientation. Its seat is the self-exceeding aspect of human nature. Man cannot remain shut up within himself for ever without ceasing to be human. Human being is basically relational, not substantive or monadic. This is not to deny or dilute his self-identity. The very social aspect of human nature clearly brings out its relational identity.

His works and words are clear evidence of his self-transcending nature. Man works in order to have what he does not have in and for him at the moment. Work is not only an acquisitive enterprise but also an act of going beyond. Similarly, his words are not merely expressive of his self but also communicative of his needs. A sense of want or negation within himself or around him is often found to be responsible for his words, written or oral. Even in his writing man goes out of himself, creates and leaves something both for himself and others. His creation takes him out of himself. This self-exceeding exercise is more clearly evident in his speech acts. His interesting words are ideological in most cases. Even monologue is a simulated dialogue. Since words and works are silently addressed to and oriented by others, their needs and expectations, our use of language is objectively required to be not only meaningful but also communicatively successful. It is in and through meaningful acts of communication that we enlarge our world, get the needed feed-back from the world around us.

Both our works and words silently follow some rules of ethics. Otherwise the very purpose of performing works and uttering words gets defeated and, what is worse, invites mute criticism and condemnation from others.

When the rules of work ethics are codified, mechanically recorded and circulated in a planned way, the whole exercise becomes technological. Our words of mouth, reduced into recorded words, are once removed from ourselves. When the same recorded words are played to others in order to influence them, the same are twice removed from us. It is in this way that our living and personal words are technologisted and step-by-step removed from ourselves through institutional mediation. The acts of self-transcendence are given a mechanical long hand. The longer the hand becomes, the reach of our responsibility, though formally extends, actually gets reduced in terms of efficacy. It is no surprise that words of mouth are ordinarily taken to be more authentic than recorded voice. That is why we hear so often the cry of primacy of perception and craze for hearing the words from "the horse's own mouth".

Undoubtedly, technology takes us beyond our narrow confines. In the process it raises some ethical issues. To what extent we can delegate our responsibility through extensive use of technology? Delegation of responsibility is like technological enlargement of our participation is collective enterprises, communication network and cognitive capacities in general, and manual and intellectual works in particular. In a primitive society where the affairs of man's life were largely manageable by his own individual efforts, his life was relatively simple and self-reliant. I say "relatively" because it is always a matter of degree. To be self-reliant is somewhat easy in a simply structured society wherein the principle of division of labour has not been taken to its extremities. In contrast, in a complex modern society, man is objectively obliged to delegate many of his own works to other human beings or institutions.

The relationship between the principal and his agent, for example, has of late assumed an unprecedented character. Our life has become so complex and the demands on our time are so numerous that we are forced to hand over or delegate more and more of our work responsibilities to such specialised agencies like the merchant banker, the cargo handler, the estate agent, the garage owner, the travel agent, the commission agent, etc. The point to be borne in mind here is that, in spite of our many technological aids of excellent and powerful type, we are required to work through delegation, discharge our responsibilities through our representatives. If I, as principal, delegate my responsibility to my agent or representative, legally speaking, my responsibility remains with myself despite the limited responsibility of my agent or representative. But when the chain of delegation becomes very long, from the principal to the agent, from the agent to the sub-agent and so on, at every step perhaps affectivity of the delegated power or authority gets gradually decreased. The legal and moral distance between the principal and the sub-agent or that between the producer and the retailer cannot be entirely bridged over to the satisfaction of all concerned. Neither technological means nor legal provisions can totally annihilate this social and power distance.

The question may be viewed from the other end. Are we becoming more and more delegation-prone because of the highly organised legal and technological character of the modern society? Or, is it due to endless profusion or multiplication of our individual psycho-somatic needs that we, in our bid to satisfy the same, are setting up or innovating new institutions and new agencies of increasingly complex structure? In brief, how are we to understand the relation between increasing needs and increasing delegation of responsibility? Must their relation be viciously circular? Can't we break the circle or make it at least a virtuous one?

The moral issue underlying these questions may be formulated in this way. When as the principal I take some morally responsible works and authorise others to execute them on my behalf, I am not on holiday from my obligation. At the same time, often it so happens that I cannot be sure either that my man will be doing the work as conscientiously as I myself would have liked it to be done. Enlargement of our area of responsibility, either due to our own needs or due to institutional and technological forces or both, seems to have made us more or less self-estranged. The distance between myself as the principal and the same self as (legally and morally) present in the authorised agent or representative can hardly be absolutely harmonious. The reason is simple. Legal stipulation and moral obligation are not strong enough forces to unify completely any two distinctly free persons. The freedom that underlies the profusion of our needs, taking us beyond ourselves, is also the only basic power available with us to control those needs. When in place of self-control we start relying on institutional controls, technological remedies, not only we ourselves become weakened but also our intention to be self-reliant gets compromised.[13]

IX

Technological self-enlargement is of two different types. And in each type the role of technology is different. It is bound to be so mainly for two reasons, the composition of the concerned culture and the stage of the concerned technological development. These apart, the constant factor, human nature itself, also undergoes historical change, at times slow and at times fast.

Man feels enlarged when his possession, material or proprietary, increases. A feudal man, for example, owning a vast stretch of land and a large cattle population thinks himself somewhat identified with it. It is to be remembered by the modern man that, though the form of life in a feudal or tribal society was relatively small in size, the sense of identity of a man at that stage was more comprehensive and less fragmentary than we see in our own time. Strictly speaking, the sense of enlarged self-identity or the narrow one is not a function of time. It is primarily a function of social structure and accompanying technology. In the ancient India, for example, there was an idea prevalent among the people that everything is available within it and that what is not to be found in it cannot be found anywhere else. The people's implicit assumption was that the larger India of the time was a world by itself. Similar ideas are found in the oral tradition of the earlier Hellenic world. About the very geographical size of the world the ideas of the proto-historic people were "quite strange" from our standpoint.

Our standpoint and position in the world have been largely influenced, among others, by fast transport system, telecommunication system and increasing migration and immigration. Today one's possession is not primarily determined or assessed in terms of one's land holding and ownership of cattle. It is determined mainly in terms of the value of one's tax-paid wealth. One's economic position is also determined by one's creditworthiness. An intelligent economist of the modern time is more interested in investment rather than in savings. He does not mind borrowing provided he is not landed in ruinous speculation and gambling. Money market of the modern man, unlike that of his medieval or ancient counterpart, is not local or even national. The emergence of transnational companies, international capital market and financial institutions has given new turn to the concept of possession and that of acquisition. Because of the changing character of technology the concept of market has been undergoing steady change. There were some forms of market in the past which could be seen, where people could assemble, buy and sell their goods and services. Still these forms of market are available in the present-day society. But there are other forms of market like international capital market or international bullion market in which some people could operate but nobody can see or visit it. It is a sort of theoretical entity which captures a number of related operations in an idealised form. These operations have been made possible largely because of efficient and fast information system, enlarged economic interests of the people and the modern transport facilities. Without the aids of modern technology international market operations would have been impossible. The invisible and conceptual nature of this sort of markets seems to clash with the concept of market defined in terms of some experienceable place, persons and operations. Enlargement of what we call market over space and time has imparted definitely new character to it. Like the impact of invisible device, for example, the impact of this sort of market on the concerned people's life, though very understandable, is largely uncontrollable. For one individual person, or group of persons or even a country, unless it is economically very strong, it is difficult to influence this market. It is true that this market can be more or less manipulated by mighty transnational companies, central banks, etc. Ordinary operators of the market are substantially subject to its forces. The only freedom they retain or enjoy in the market is *not* to operate in it when the going are very bad for them. But this is true only up to a point. Because an international currency dealer, for example, cannot withdraw from it for ever without harming his own professional interest.

Self-enlargement in terms of international market entails a sort of self-estrangement. In a lesser way every person, buyer, seller or agent, who operates in a market is bound to be determined by the forces of market. The rules of the game have to be obeyed. Unruly behaviour, behaviour inconsistent with the rules of the market, invites punishment, i.e., loss.

Self-enlargement is achieved also in terms of love and sympathy. The pacifist, the ecologist, the artist and other gifted persons who try to share their experiences, aesthetic, ethical and creative, are also engaged in making our world larger and larger. The writings of the great poets, the great novelists and the works of the great scientists and technologists contribute much to the cause of globalisation of our consciousness. This may not be their primary conscious aim but this turns out to be a welcome consequence of their own creation and innovation.

Enlargement of freedom through creative works implies least alienation or estrangement. Because the appeal of the works of the poet, the painter and the scientist, for example, are not intended, certainly not primarily, to encroach upon others freedom. Positively speaking, the main appeal of their works lies in freeing human sensibility itself from the customary forms of expectations. By experience of what is true, intellectually satisfying, what is beautiful and what its morally elevating our consciousness expands. These nobler modes of experience make us citizens of an expanding world. It is true that to make this modes of experience possible we do not directly make use of any technology as ordinarily understood. However, closer scrutiny reveals that none of the creative modes of experience is absolutely spontaneous. To be scientist one has to undertake serious research work, theoretical or experimental. Even theoretical research at some point or other is in need of some or other technological aid for validating its results. One cannot be great poet merely by the grace of divine intoxication. One has to work hard and that too for a long time for the purpose. Besides, one needs to master some style and technique in working out his themes. This is true not only for the poet but also other forms of artists like the musician, the painter and the sculptor. The artist uses, though inarticulately, some or other forms of technique.

In expanding the world of arts and science the technology which has played the most major role is printing. Arts in oral tradition, science in practice and some or other sort of technology in use were there long before they have been given institutional forms. The question of reducing folk culture in words (literature) or colours (picture) did not assume a significant character before the invention of printing technology. Possibility of mass circulation and availability of art forms became real only when printing in general and modern forms of printing in particular became current.

Here also one must perceive the distinction between direct or face-to-face experience of the same in books and other recorded devices. To learn science from the courses offered on TV by Open University and from the lectures of the class-room teachers and the demonstrators in laboratory are considerably different. What is said in this context in Indian tradition is noteworthy. The most profound forms of knowledge and artistic excellence are to be acquired only through direct or face-to face relation and not through institution. The knowledge of ultimate reality *(Brahmavidya)* for example, cannot be acquired except through directly from the teacher *(gurugamya vidya)*. *Darsan* (philosophy), *yoga*, music and dance are among the most important forms of *gurugamya vidya*. The idea is that whenever knowledge or excellence is institutionalised and imparted through the concerned institution, and *not* directly through teachers, there is bound to be some loss of quality in the process of communication or transmission.

Here, once again, we find the analogy between the role of institution and that of technology. Their common denominator is impersonalisation or objectification of knowledge and skill. If estrangement is due to institution and technology, it is bound to be acute in those institutions which are unlived or like shell without substance and in those technologies which have no human face and touch.[14]

The negative sides of technology and institution must not make one think that these are totally dysfunctional. But for them our self-encounter would have been completely free. This is a wrong idea. Institutions would not have been there at all if they were of no use for human beings. In a way they are expressive of human nature and needs. This is the point I have tried to emphasise by denying the possibility of reduction of societal facts and institutions into individual, their dispositions and actions. Similarly, technology is not be taken as an accidentally received device. it is a human innovation and its root lies in human consciousness and needs. To try to understand technology exclusively in the analogy of an external adjunct is quite misleading. As I have said earlier, it is an expression and extension of our very being.

At the same time it has to be admitted that enabling conditions of our expression, paradoxically enough, prove to be obstructive of expression under some specific circumstances. The classical example of this point is our own body. If I say "I am *not* my body", I am partly right. But not entirely. On the other hand, if I say "I am not without my body", I am also right only to certain extent. And not entirely. My existence is embodied. My freedom needs locus, is not disembodied, and, positively speaking, it is in and through my body that my freedom is articulable. That the forms of articulation of our freedom are found to be numerous is again partly due to body, different senses and organs endowed with specific qualities. This is not to deny that even in consciousness there are different forms or modes, ethical, aesthetic, institutional, etc. The main point I emphasise here is that right in our own body we have both an articulative and a restrictive condition of our freedom.

More than body what restricts our freedom is others' freedom. Freedom *qua* freedom is obviously not divisive or conflict-inducive in character. It is institutionally mediated nature of freedom which is largely responsible for inter-individual, inter-institutional and inter-cultural conflict. In their efforts, honest or otherwise, the individuals who man different institutions and shape different cultures try to enlarge the sphere of their freedom. In the process often they disregard the sanctity and legitimacy of others' spheres. Institutions and cultures are put to conflicting relation mainly because of the conflicting interests and aims of the persons concerned.

Institutionally mediated conflict is intimately related to technology, for example, the technology suitable for small-scale industries, unless protected by appropriate legislation, is likely to be swept away by the mighty technology of the large-scale industries. Similarly, the interests and aims of Employers' Association, except under highly ideal conditions, do not prove harmonious with those of Trade Union. Democracy or freedom as such is not enough to ensure their smooth relationship. On the contrary, in the name of market freedom and also in the name of democratic Trade Union rights, inter-group conflict tends to become intense. In the name of cost-efficiency producers are often inclined to reduce the privileges of the employees and the work conditions are allowed to be relatively unattractive, if not unwholesome. For the resolution of these conflicts different mechanisms like negotiation, arbitration and adjudication are undoubtedly available. This availability is primarily notional and expressive of the freedom enjoyed by the society as a whole. But, interestingly enough, it is mainly within the free society that the causes of conflict are not only allowed to flourish but also given recognised forms. In other words, institutions of the free society, aided by technology, make conflict an integral part of the freedom of society.

At the private level it is again in the name of freedom that privacy is encroached upon. Similarly, freedom-related privileges like anonymity are also substantially curbed in free society. Technology is largely responsible for curtailment of our privacy and anonymity. For example, my telephone, though a facility for me to be connected with the persons I need for myself, exposes me to uninvited and at times even unwelcome phone calls. I may be called at odd hours. Besides, my number and address given in the telephone directory create many problems for me and for family members. The point I am raising here may be sought to be rebutted in this way. For example, one may point out that if I for my privacy refuse to make my number and address public and others also propose to follow the same path, there is likely to be chaos, break down of communication, if not of the public life itself. If the ways and means of availing utilities are all kept secret, the quality of social life, the use-value of social services are bound to suffer. Privacy and anonymity do not necessarily require to be protected by secrecy. In the mass society of the modern time it is very difficult to enjoy undisturbed privacy. Anonymity is also subject to related difficulty. Freedom is bound to have some disturbing or even unforeseen aspects of itself. But that is not a good enough ground to restrict freedom unduly. Because restrictive freedom is destined to create larger problems of more serious type.

The problems of freedom may be highlighted or even dramatised by referring to the roles of institutions like prison and asylum. Many prisoners honestly believe that their imprisonment is due to conspiracy of others and failure of police and justice. Many inmates of asylums are heard to observe that they are fully sane and that they have been brought to asylum by the unloving or hostile members of their family and with the connivance of the unqualified or unethical psychiatrist. Even if we do not take the self-justifying prisoners and inmates of asylums in their face value, perhaps the point remains that at least some of them do not deserve to be where they have been put by the circumstances of life or miscarriage of justice.

Whether technology has a say in curbing human freedom or in removing the causes of the curtailment of freedom? If so, to what extent? In search of valid answer to this question we are required to look into technological underpinnings of the modern society. Why it is so? More and more people in the industrialised society are getting mentally ill. Why the big cities of the industrialised society are found to be more crime-prone than the cities and particularly villages in the less industrialised society? One stock answer to this sort of questions is along this line. The conditions under which the cases of mental illness and crime are found out and made public are just not available in the less industrialised society. The apparent difference between the two social scenes is attributed to the difference in the facilities of diagnosis and detection and not in the substance of the matter itself. This response may be partly correct. But this does *not* explain the whole of the said difference. In order to understand the situation more satisfactorily we have to look closely into the estrangement and dehumanising role of technology in the industrialised society.

The main problem of the technological society seems to be its perpetual tantalising character. It remains more tantalising than satisfying. Profusion of needs gives rise to greeds and is reinforced by the same. Under the impact of unsatisfying and diversifying needs, our life gets increasingly directionless and fragmented. When one is not sure of what exactly one wants one cannot be contended in life. When one's focus of interest starts constantly shifting because of alien or unlived interest, one is treading a slippery path. The fast pace of social change itself experts an unsettling influence on our forms of life, familial, professional, political etc.

Uneven levels of achievements, abilities and aims of life of the members of a family creates a tension within the family itself. What adds to this tension is the social proximity of other pulverized families. The break down of family life need not be always structurally manifest. Often it takes the form of what may be called schism in the soul. When both the parents work, the children are bound to be more or less emotionally neglected. Those who have not enjoyed love, affection or emotional attachment in their own life are not generally found to be very helpful or warm to others. But for the emotional neglect of the children the parents cannot be entirely blamed. For pressing economic needs, real or imaginary, they want to increase their income both for themselves and also for the children's comfort and security. Implicit in this dominant economic attitude towards life is the persistent sense of insecurity. The persons who are anxious of their own future and afraid too much of imponderable cannot form a steady frame of life for their ownselves or for their children. Objective study shows that most of the forceful imponderables are unfounded or imaginary. But the members of the technological society under the constant pressure of what they hear and see, factual and fictional, feel threatened by the forces beyond their control. Their self-identity becomes opaque or more or less incredible to themselves. They are seized of by a sense of rootless anxiety about themselves. Consequently their concern for others goes down and even their self-esteem gets eroded.

X

The ambivalent character of technology has been indicated in the very title of this paper. Technology as technology is neither promoter nor destroyer of culture. The genuine critique of technology must not paint it either as very admirable or very despicable. Its context and culture-specific character has to be brought out so that the true role of it could be clearly appreciated.

As I have tried to indicate earlier, technology is to be understood primarily in its relation to human labour, manual and intellectual. Further, labour is culturally embedded. Therefore, in the process of evaluating technology, we are obliged to take note of the relation between labour and its embedding culture. Is culture, including its technology, rooted in labour? The affirmative Marxist answer to this question is in a way somewhat narrow. For, it highlights the role of manual labour and rightly laments over its undervaluation in capitalist culture. It is to be remembered that culture owes its life-breath not only to manual labour but also to intellectual labour. Both works and words contribute to the rise (and also fall) of culture. Their value and efficacy are not intrinsic to them. Much depends on how they are used, by whom and in what context.

Similar ambivalence is associated with alienation. It may be depicted as a harmful and many-sided social phenomenon. Certainly in the age of capitalism and due to intensive application of technology the fabric of human relationship is getting increasingly fragmented. In the modern age man is certainly deeply afflicted by emptiness, loneliness and a sense of meaningless of all that he does and achieves in life. But this is not the only picture of alienation. First, it has its positive aspect, as distinguished from its (negatively) normative or pejorative, aspect. We are alienated not merely because of machination of technology. It is partly due to our own option. We have consciously opted for technology. Having exercised this option, we cannot honestly attribute its negative after-effects to technology itself. Secondly, technology is backed up by increase of institutional mediation. Multiplicity of institutions and increase in division of labour are partly responsible for the blame that we often uncritically lay at the door of technology.

The problem of alienation, for example, deserves to be treated more positively or descriptively than we have seen mainly under the influence of the Marxist analysis of the phenomenon in the early capitalist culture. The accent on technological alienation or estrangement is mainly due to our preoccupation with its productive and exploitive aspects. But, rightly understood, technology has a place for it in our very thought-process. Increasing complexification and conceptualisation of our thought make use of some logical, graphical and mathematical techniques and as a result of that we are simultaneously cramped as well as facilitated in our consciousness and thought exploration of the world around and within us. The role of technology in our life is determined to a certain extent by our consciousness of it and by the role we wilfully assign to it. In other words, the autonomy claim or the "Frankenstein" character of technology cannot be accepted without serious qualification.

The point may be clarified in this way. Neither as worker nor as thinker we are entirely at the mercy of our technology. Are we not ourselves choosing our technology in many cases? In order to perform some of our works in a better and acceptable way we change our technology. Technology adds value to what we produce by our hands or use of head. To add value to our own work, to be influential in life and also to be more competitive in market we keep on changing our technology. It is true that in some cases, in large-scale production units, for example, the workers ordinarily do not choose their technology. We are not always master of our technology. But our subjugation, especially our sense of subjugation, to technology is certainly influenceable by our own thought and action to a significant extent.

Whatever may be the character of the institutions to which we are affiliated and whatever might be the size and level of the technology we make use of, our responsibility as thinker or worker or both is never negligible. Responsibility cannot be wholly dictated from without. Neither the owner nor the manager of a workshop can entirely control the nature of work we are called upon to perform. For earning our bread we may sell our labour. But our soul is not on sale. It is often for the butter on the bread, our "extra" need, that our soul is put on sale. This important distinction between bread with butter and bread without butter is neither rhetoric nor metaphoric. Certainly not in the context of poor and developing countries[15].

Freedom cannot be easily crushed by others unless we silently acquiesce to this exercise. The relation between our freedom and what negates this freedom are both rooted in our own self-existence. We are both producer and consumer of our freedom. The pattern of our production and that of consumption cannot certainly be determined entirely by ourselves. Nor would it be correct to assert that these patterns are wholly imposed upon us by our attending institutions and technology. here lies the perilous as well as promising field of our responsibility. Forms of technology come and go. In most cases these are not renewable. But the needs and the mind underlying those needs which invite new technology and reject the obsolete one maintain their identity and continuity in intimate and identifiable way. This suggests that as workers and thinkers we are at an advantage *vis-a-vis* our technology and institutions. We can control them more effectively than they can possibly control us. Here seems to lie the key to proper understanding of man-technology relationship.

Notes and References

1. David de Hahn, *Antique Household Gadgets and Appliances, Baron's Woodbury, NY*, 1977
2. R.L. Gregory, *Eye and Brain*, Weidenfeld & Nicolson, London, 1966.
3. Joshua Meyrowitz, *No Sense of Place*, Oxford University Press, Oxford, 1985.
4. George McRobie, *Small is Possible*, Harper & Row, New York, 1981.
5. Bernard Jendron, *Technology and the Human Condition*, St. Martin's, New York, 1977.
6. D. Lovekin, & D. Verene, eds., *Essays in Humanity and Technology*, Sauk Valley, Dixon College, Illinois, 1977.
7. Kostas Axelos, *Alienation, Prasix and Techne in the Thought of Karl Marx*, University of Texas Press, Austin/London, 1976.
8. Robert Hugh Macmillan, *Automation, Friend or Foe?* Cambridge University Press, Cambridge, 1956.
9. E.F. Schumacher, *Small is Beautiful: Economics as if People Mattered*, Harper & Row, New York, 1973.

10. F. David Noble, *Forces of Production: A Social History of Industrial Automation*, Knopf, New York, 1984.
11. Lewis Mumford, *The Myth of the Machine: Technics of and Human Development*, Harcourt Brace Jovanovich, New York, 1967. See also, Cohen, Marshall, Nagel, Thomas and Scanlon, Thomas, eds, *Equality and Preferential Treatment*. Princeton University Press, Princeton, NJ, 1976.
12. Robert J. Miller, ed., *Religious Ferment in Asia*, University of Kansas Press, Lawsrenece, KS, 1974 see also Berle, Adolf A., *The Modern Corporation and Private Property*, Harcourt, Brace and Wrold, New York, 1968; and Nathan D. Grundstein, *The Managerial Kant: The Kant Critiques and the Managerial Order*, Case Western Reserve University Press, Cleveland, 1981.
13. Brent Fisse, and Peter A. French, eds., *Corrigible Corporations and Unruly Laws*, Trinity University Press, San Antonio, TX, 1985; see also, Werhane, Patricia, Persons, Rights and Corporations, Prentice-Hall, Englewood Cliffs, NJ, 1985.
14. David Macarov, *Work and Welfare: The Unholy Alliance*, Sage, London/Beverly Hills, 1980; see also, Arendt, Hanna, *The Human Condition*, Doubleday, Garden City, NY, 1959.
15. D.P. Chattopadhyaya, *Sri Aurobindo and Karl Marx: Integral Sociology and Dilectical Sociology*, Motilal Banarsidass, Delhi, 1988.

Biotechnology and Ecological Security

M. S. Swaminathan

Ecological security, which involves the conservation and sustainable management of our basic life support systems consisting of land, water, flora, fauna and the atmosphere, is basic to both food and economic security. To achieve ecological security, we need to fulfil five basic requisites.

They are:

a. Stable human and animal populations.
b. Development and adoption of environmentally benign or "green" technologies in industry, agriculture, energy, transportation and other systems.
c. Removal of absolute poverty through opportunities for work and minimum income for all.
d. Environmental accounting procedures which consider natural resource deterioration as another form of capital depreciation, and
e. Environmental literacy which generates awareness that the protection of the environment and prevention of pollution is the first and foremost duty of every citizen.

By the year 2000, nearly 70 percent of our population will be below the age of 30. *While the famine of youth will be a serious problem in industrialized nations, famine of jobs for youth will be our most serious problem.* Without adequate opportunities for skilled employment, poverty will persist, urban slums will grow and social unrest will undermine political stability. It is in this context that recent developments in biotechnology assume significance.

The hardcore of biotechnology is molecular biology and genetic engineering. The emerging tools for the transfer of genes across sexual barriers provide unusual opportunities for promoting sustainable advances in the productivity of crop plants and farm animals. This is essential, since both land and water are shrinking resources for agriculture. The other major applications include the use of tissue culture techniques for propagating superior material of horticultural plants and forest trees, diagnosis and control of animal and plant diseases and desirable biological transformations induced by micro-organisms. In addition, new opportunities exist for the preparation of value-added products from every component of plant and animal biomass.

Biomass Refineries can be established for this purpose. *Biotechnology is a human resource and plant and animal germ plasm based industry.* In both these resources we are rich. It will be foolish on our part not to derive benefit from these twin blessings — our vast population and our rich biological wealth. What we need for this industry are large number of young women and men who have passed their 10+2 examination and who are willing *both to unlearn and learn.*

The potential benefits from the use of genetically improved plants and micro-organisms are enormous. We may give priority to fruit and forest trees that grow more rapidly, bacteria that biodegrade environmental pollutants, viruses that control insect pests, vaccines for animal diseases, food plants that grow well under saline or dry conditions and to material tailored to the needs of international markets. At the same time, we should develop and enforce stringent safeguards which will prevent possible hazards to human health and damage to the environment from transgenic organisms. We should also ensure that discussions in International forum like GATT do not lead to an enlargement of the scope of Intellectual Property Rights to include genes and breeding material. We should while fighting against the extension of the scope of rewards provided to the formal innovation system, *develop a proposal for according recognition to the informal innovation system based on traditional know-how and material.* In agriculture, formal innovation yields maximum benefit only when it is based on the foundation of the informal innovation system.

Ecological security can be ensured only by eliminating the triple threats arising from population, poverty and pollution. Employment security and environmental literacy are both essential for minimizing these triple threats. Biotechnology promotes opportunities for skilled jobs and for developing awareness through action programs. It also provides an opportunity for achieving sustainable advances in biological productivity under conditions of shrinking land and water resources and expanding biotic and abiotic stresses.

The tools for biotechnology should also be used for preparing ourselves to meet potential changes in climatic condition, such as altered precipitation and temperature patterns, sea levels and incidence of ultraviolet radiation. Under the auspices of my Research Institute (Center for Research on Sustainable Agriculture & Rural Development) in Madras, we are establishing a *Genetic Resources Center for Adaptation to Climatic Changes*. The aim of this Center is to assemble genetic material capable of providing donor genes for tolerance to drought, salinity and sea water intrusion. Such anticipatory research is also an essential component of a national ecological security system.

An important contribution of biotechnology to small farm agriculture is the opportunity it provides to add the dimension of sustainability to land use systems. Hence, I wish to deal briefly with certain aspects of sustainable land use. Due to the growing pressure of population on land, we will have to produce more and more food and other commodities from less and less land. Can sustainable advances in biological productivity be achieved to meet the needs of the growing population? The answer is "yes", provided several steps are taken, among which the following are some of the more important.

a. A method of measuring productivity should be introduced, which permits quantification of the impact of productivity advances on environmental capital like land and water.
b. Methods of environmental accounting which help to measure sustainable income by *taking into account costs of damage correction* (such as irrigation induced salinity, etc.) *and capital degradation, should be*

developed in crop husbandry, animal husbandry, fisheries and forestry.

c. A sustainable food and nutrition security compact giving ground rules for the equitable and ecologically sound use of the basic agricultural assets should be developed.

d. *Equity and location specificity* are two important characteristics of the pathways to sustainable development. Hence, the capacity for location-specific technology development jointly with farming and fishermen communities should be strengthened.

e. In recent years, energy equipment have increased by over 2 times for every unit of output in agriculture. Energy use efficiency should hence receive priority attention.

f. The sustainability aspects of land and water management, crop and pest management, energy management and post-harvest management need analysis both on a factoral and systems basis.

g. *Sustainability should be regarded as a dynamic concept.* Indicators of sustainable development should take into account this dynamic aspect. Unacceptable risks in areas such as soil productivity, water quality and availability, biological diversity, pollution and energy sources should be defined for each agro-ecological zone. Indicators of unsustainability should be developed.

h. A sustainable aquifer management policy and an equitable water sharing policy are particularly important in semi-arid regions. Where there is no equity in water sharing, there will be no co-operation in water saving. Hence, the growth of conservation minded water shed communities should be fostered in such areas through appropriate public policies such as group insurance and group extension.

i. Sustainability can be achieved only through the application of the best in modern technologies, such as biotechnology, space technology, information technology and management technology, to promote the growth of conservation-based farming techniques. For this purpose, there is need for appropriate research, extension and training efforts. Women, in particular, could be trained to take a leading role in the conservation of biological diversity, integrated pest management, organic-recycling and use of bio-fertilizers, energy management and post-harvest technology.

j. Farming Systems Research (FSR) methodology should be re-designed as a Rural Systems Research (RSR) program in order to link the primary, secondary and tertiary sectors of economic activity in a mutually re-enforcing manner. If this is not done, the pressure of population on land will increase and it will be difficult to adopt the techniques of land saving agriculture and grain saving animal husbandry.

We have no option except to produce more and more food from less and less land. The carrying capacity of land has to be a dynamic concept, although it cannot be stretched beyond limit. Our aim should be to develop and disseminate "green" or environmentally friendly agricultural technologies which can help to optimize employment, income and agricultural productivity from units of land, water, energy, labor and credit. The social engineering and organizational aspects of sustainable agriculture need greater attention. This will call for the development for every village or a cluster of villages *a sustainable agriculture matrix.* Such a matrix should show the implications of the imperatives of ecology, economics and equity for the packages of technology services and public policies developed for widespread adoption.

Land is the principal source of both food and livelihood security. We should foster the *establishment in every Block of the Country a Sustainable Land and Water Use Board,* chaired by an active farmer which can help to promote the use of land in accordance with the following three needs:

a. *Conservation areas,* which should be maintained in their pristine purity.
b. *Restoration areas*, which are usually referred to as "wastelands" whose biological potential has to be upgraded through the principles of restoration ecology, and
c. *Sustainable intensification areas,* where productivity can be continuously improved through attention to soil health care and plant protection.

The nineties will be a critical decade in our agricultural history. The future of agriculture will depend upon our ability to promote sustainable land and water use systems.

Unless we develop suitable institutional structures like those outlined in the paper, it will be difficult to harness the tools of biotechnology for promoting ecological security.

*Non-Alignment, Self Reliance and Innovation**

Ashok Parthasarathi**

The coupling between self-reliance, non-alignment and innovation in science and technology is an interesting coupling full of possibilities. Self-reliance is to be understood in terms of autonomy: autonomy for a society to define a preferred future for itself; autonomy in terms of decision-making, autonomy in terms of defining the character and composition of the scientific and technological (or research and development) effort including its scale, orientation and thrust. When defined in this manner one can see how self-reliance is complementary to non alignment as enunciated by Jawaharlal Nehru, for non-alignment basically asserts a nation's right to autonomy in international relations and in shaping a preferred world order. It has taken many

* *The paper is based on the presentation made at the Seminar on "Science in Society" held on the occasion of the Cochin Science Congress, February 1990.*

** *The views expressed in this paper are the personal views of the author and do not necessarily reflect the views of the organisation to which he belongs.*

industrialised countries a long time to appreciate that non-alignment and self-reliance are natural concerns of newly independent, ex-colonial societies seeking at least some degree of "elbow room" for themselves in the new world order of the second half of the twentieth century. It has become fashionable in some quarters in North America and Western Europe to argue that developments in Eastern Europe and the Soviet Union over the last few years have made non-alignment merely a historical phenomenon and that the world order does not call for it any more because bi-polarity has ceased. The validity of such a view, however, is quite doubtful. And so far as the legitimacy of self-reliance is concerned, this too is under constant assault both from within our society and from outside; from sections of both the natural and social science communities and from other actors in the policy-making process, including political, business and media groups.

The character of science in the post-1945 world is distinctly different from that of even the first half of the 20th century, let alone the 19th century. The pre-1945 scene was characterised by the lone scientist or at most small groups of scientists working in universities such as Gotingen, Cambridge, or Copenhagen. Their main task was teaching; research was a subsidiary activity. The principal justification for being paid was transmission of knowledge to the next generation. The creation of that knowledge was something for which they were not accountable to any public entity either in terms of financial resources or in relation to the achievement of social goals. The post-1945 world on the other hand has seen science emerge as a vocation. Scientific research is now undertaken by teams if not task forces. Whether the task is the bringing about of the green revolution or understanding the cosmos, the research effort is mounted and executed on a near military basis with a very definite command structure, both managerial and intellectual among the participating scientists. This has significantly altered the nature of science and the ways in which it has grown.

The second feature which has a significant bearing on both the drive for self-reliance and the nature of the process of innovation in S&T, is that such innovations are now heavily influenced by the policy instruments of the State even in the most market-driven economies. For example, even in the USA not only is 46.1 percent of the total national R&D budget directly financed by Government, but the Government also influences the orientation and thrust of much of the remaining 53.9 percent through the R&D needs of the Defense, Aerospace, Atomic Energy, Oceanographic and other government programmes. So, the State is, in fact, determining to a significant extent; the composition, quantum and direction of the advancement of science. When and in which area scientific breakthroughs will occur is no doubt always determined by the quality of intellectual input, but this is now matched by the institutional, policy, financial and other large-scale social structures defined by the State. The technological preparedness of the OECD countries to have substitutes to oil in 1973 provides a good example. Up to 1973, the scientific basis for utilising coal, let alone renewable energy sources, was woefully inadequate, whereas the geology, the geophysics, the seismology and, of course, the chemistry of petroleum exploration, extraction and utilisation was highly developed. This was not an accident. It was in fact targeted, driven by the State alongwith transnational enterprises. One can see the beginning of a similar phenomenon already emerging in as recent a field as biotechnology, where, because of the concern of the industrialised countries with medical applications, medical biotechnology appears to be growing much faster in terms of basic scientific work than biotechnological innovations for industrial applications. In India, too, the character of basic scientific research in biological sciences is influenced by this trend. There are, no doubt, additional reasons arising out of the desire to substitute available pharmaceutical products by new bio-tech pharmaceuticals because of enhanced awareness of toxicity, safety and so on. The pertinent point to note is that the scientific research effort is being determined not necessarily within the framework of a predetermined body of knowledge being unfolded by the process of scientific discovery (which is what we identify scientific research to be), but more by demand considerations generated by State policy. The direction of scientific work, the kind of relationships between basic science, applied science, and development, even before coming to technology, is strongly determined by the macro-orientation of State policy. The relative positions of biological and physical sciences in our country is a classic manifestation of the manner in which State policy has shaped the orientation and composition of scientific advance. Despite our obviously much greater social need for biological scientific knowledge of our flora and fauna and of the technology development based on that knowledge, because much of our economic activity is based on the bio-mass infrastructure, the bulk of our scientific research effort is still in the areas of physical sciences. The third feature of the post-1945 scenario relates to technology. There is now wide spread awareness in the scientific community, as also amongst the technological community, that science and technology are different; they have differences in origin, character and dynamics. While it is true that S&T intersect at many points, the intersections becoming less and less matters of accident and more and more matters of design in terms of policies, programmes and institutions.

These three features enable us to comprehend the significance of self-reliance. From the point of view of the autonomy aspect of self-reliance, this changed scenario of science & technology gives us expanded opportunity for creative and adaptive responses in terms of a more policy-oriented view. The autonomy question is, in fact, the central one. Looked at from that point of view, I think, we are now in a far better position to determine what should be the mix in the 'S' (science) segment in terms of disciplines and trans-disciplinary bodies of knowledge as also what should be the mix in relation to 'T' (technology) as determined by the pressures and reactions from economy and society. Those pressures and reactions first come on 'T'. The hybrid seed is, in fact, an instrument of 'T'; it is not an instrument of 'S'. Keeping in mind the pressures from the society and the economy that come through the 'T' process, how does one link 'T' and 'S' when such linkages are called for?

A particularly important issue relates to determining whether, and if so, to what extent, there is really need for scientists to involve themselves in the question of choice, i.e. where do we put resources, how do we deploy resources of all kinds — human, physical and infrastructural? What are the criteria of choice between fields? Should they be driven completely backward by the economy and society impinging on 'T' with 'T' being the customer of 'S'? When we do 'S', should 'T' be taken into account? When Marx coined the phrase 'Science as a direct productive force in the economy', 'T' was really not known. Without the changes that have taken place since 1945, 'T' would continue to be treated by economists as 'a residual factor' in economic growth. But, we now know substantially more about how 'T' works.

Another important point to note is that (as shown by studies conducted by the Nomura Research Institute, one of the largest policy think-tanks of Japan) there have been a very large number of important technological breakthroughs in the post-1945 world. I am deliberately using the word "technological breakthroughs" so as to exclude the small changes and improvements in technology brought about on a cumulative basis mostly in the process of production itself. Such technological breakthroughs have come more from the 'T' of the previous generation rather than from the 'S' of the new generation, a view scientists may feel uncomfortable with but which is now well documented in the literature.

One is not suggesting that the Japanese experience can be directly extrapolated to the Indian and other situations. There are a number of differences — our extensive agricultural and rural base, the composition of our economy, our work force, etc. So, one-to-one type of association is not possible. For example, Robert Solow's book, *Manufactured in America*, is in a certain sense an answer to Akio Morita's book *Made in Japan*. *Manufactured in America* says that if at all the United States is to regain its healthy economic position both domestically and internationally, the one thing it should not do is to copy the Japanese model because of the numerous and large differences between the two societies. But Solow also makes the point that a lot of 'T' for breakthroughs is required. Developent of 'T' not in terms of cumulative small changes at plant or enterprise level close to the production front but major technological innovations such as the Transistor. However, in our situation the area of such small cumulative changes has been significantly ignored at very significant economic cost. Our concern that we should sharpen the initial tool with which we start an industrial enterprise by improvements to it made by practitioners of the production process has been practically nil. What is the technological environment inside the enterprise? What is it in the market? What is it in terms of state policy that has so inhibited or atleast not stimulated such cumulative technological improvements. These and many other issues are involved in the whole question of small 'T' following on large 'T'.

But I would like to submit a more fundamental proposition: viz. that new 'Tl' flows from old To and new 'T2' flows from old Tl. This is not to say that 'S' will not intersect with either Tl or T2. But the unilinear model of innovation with which we have worked so far, which set a new body of knowledge to spawn a new body of technology leading to new capability in the economy and new industry, new products, new jobs and so on, is no longer a valid generalisation. There was a product cycle, the cycle varied, and we got a new 'S' curve coming through the same chain. That, I think, is no longer the case, certainly not as a generalisable proposition. Individual cases may well follow the unilinear model, particularly in nascent industries like bio-technology. But very little is as yet known about such production structures.

The last decade has seen an increased awareness in this country that innovations are to be viewed in relation to the economic enterprise, a structure which introduces goods or services into the economy, whether it is household, cottage, small scale, or large scale; whether it is agricultural or industrial. However, there is still inadequate awareness that, innovations are of many different types and that significant improvements in productivity, quality and costs do not necessarily call for innovations in physical technology. Many of the Japanese improvements have been in social technologies. At international meetings leading Japanese scientists, economists and business managers all emphasise this "soft" (institutional, information-based and behavioural) dimension of innovation. They talk about institutional forms, communication forms, hierarchical organisations, horizontal organisations, etc. There is a concentrated focus on these dimensions and a de-emphasis or deliberate playing down of the tangible, hardware dimension of technology.

In India, one of our most serious challenges is going to be: can we, again without wishing to copy from the experience of other countries, but certainly with a wish to learn, broaden the understanding of innovation in the economy, not necessarily innovation through physical technology but also innovation through social technology? Therefore, it is not just self-reliance in physical science and technology that we need, but self-reliance in terms of activities across the spectrum, with more and more emphasis at the "soft" end viz. the social and behavioural sciences. All those who have run institutions know only too well how small improvements in this "soft" end have led to tremendous multipliers in terms of output.

If one is to capsule into the 1990s and to 2000 and beyond what should be the kind of thrust with respect to self- reliance and innovation in an autonomous India maintaining its flexibility and freedom of action in world affairs as reflected in the term 'non alignment', I would say that the 'T' invasion needs to be appreciated in full to recognise that more and more new 'T' is coming out of old 'T' and not from new 'S'. That ultimately an enormous part of our existing capital assets and our knowledge capability is in the hands of those who are practitioners of technology, in industrial plants and in agricultural fields. This is not to say that those who have specialised, formalised knowledge, are not important, but we need to redress the balance between the contributions of the two. And finally, innovation is a multi-dimensional, multi-disciplinary business. It is not related only to the natural science component, but also to social science, behavioural science and social technology of organisational management, motivation and values.

I would like to conclude by referring to a statement made to me privately by the Japanese economist Prof.Hayakava. He said, "Our experience in Japan shows that the most important ingredient for the success of any institution, whether it provides goods or services or even knowledge to the economy and society, is what we call "the initial kick." How is the institution launched? What is the organisational, psychological, manning and skill structure at the time the institution is launched? He emphasised how much care Japanese institutions take, even academic institutions, in forming new departments, new centres, etc. let alone industry, in considering how that initial kick needs to be given. In India, the closest we came to the initial kick, of course, was the Bhabha formulation — 'find the man and the build the institution around him'. But there are other very different elements in the initial kick that are important, because the initial kick is a planned operation, very carefully thought through to control as many variables and to protect against many assaults that the new entrant will face. Building around a key figure is obviously a vitally necessary element, but not a sufficient one. To recognise the need for structured multi-disciplinary approaches to innovation is the real challenge.

Sustainability and the Energy Picture : India

C.V. Seshadri

Central to development is energy. More important, perhaps, is the way energy is incorporated in the plans and plan-models for development. Most plans and plan-models, however, lay emphasis on the role of money and its supply in driving the process of development and tend to describe the entire plan in largely fiscal or monetary terms. For example, in a recent official appraisal of the VII Plan, the progress of the plan is discussed primarily in monetary terms of capital accumulation, cash reserves and cash flow, investment, lending and the like[1].

The advent of the "supply-side" approaches in economics about thirty years ago resulted in frameworks of analysis which tended to neglect the impact of changes in energy supply on the overall supply of "goods and services" in an economy. Typifying such approaches, the sudden, drastic increase in the oil prices in 1973 has been appreciated as being economic in character. Further, the finiteness or depletability of many resources (particularly of fossil fuels) is not seriously considered in many of the analytical models.

In India, energy-cost is often treated as a significant parameter that could affect the overall performance of the economy, but this is considered strictly true only for what are called "commercial sources of energy". The other sources, "non-commercial sources of energy" as they are termed, are derived from the biosphere (or biomass) and as such are not assigned proper value in the economy. In the developed countries, rising economic costs of energy led to the development of conservation strategies in energy utilization (or extraction). But the lack of proper quantification of the value of biomass sources in general has been responsible for the absence of viable conservation strategies in biomass energy utilization in India.

That the ultimate source of economic value was in the natural resources (i.e. land) had been the basic tenet of one school of economic thought (called "physiocrats") in the 18th century.[2] In the last century, Jevons, an English economist, emphasized the crucial role played by the availability of coal in making Britain the leading economic power of the day, and cautioned that too rapid a rate of exploitation of that resource might have disastrous consequences for the pre-eminent economic position of England. There were several others, mainly scientists, not economists, at the turn of this century, who repeatedly emphasized the role of energy availability and its utilization in the development and progress of national economies, or even 'civilizations'.

Attention to such approaches to economics was gradually diverted since the late fifties when two types of arguments were advanced: the first, was based on the observation of time series decreases in the quantity of labour and capital required to "harvest" natural resources.[3] On this basis, it was argued that the interrelationship between energy-cost, or the costs of natural resources, and the overall economic growth could be decoupled over the coming years. The other type of argument, is based on the belief in the complete substitutability of energy by capital and labour.[4]

Among economists, Georgescu-Roegen has contested the notion of full substitutability of fuel/resource by labor and capital.[5] His point is that generation of capital goods such as machinery itself requires use of a natural resource but capital cannot create the resource out of which it is made. Thus, there obtains a new relationship between resource and capital, leading to the view that a resource is not like any other commodity governed by or subject to the action of market forces.

The nature of the energy-basis of the substitution process has been brought out by Tyner and others[6], who considered the following example: when the price of oil in the U.S.A. was $10/barrel, it was said that if the oil price increased, shale oil ($30/barrel) would become competitive when the former reached $30/barrel. However, even when this price level was reached, shale oil had not become competitive; this was because the hidden cost of the requirement of 3 barrels of oil in the extraction of 1 barrel of shale oil had been overlooked.

In a proper determination of the value of energy available from a conversion process or source, the framework developed by Odum is likely to be useful. Odum's approach consists in viewing any energy source as transformed solar energy. The value of energy derived from a source is thus estimated as the value of solar energy which it took to "embody" the source. Such estimations can be carried out using the solar transformity ratios reported by Odum.[7] The principal advantage in the use of this framework is that energy derived from the biosphere (e.g., fuelwood) and fossil fuels could be compared on a unified basis: the extent of depletion that the use of a given process may cause can also be estimated.

Value of energy: the importance of evolving a common methodology and convention

Since the last 15 years, Energy Analysis (or Net Energy Analysis, as it is called in the USA) has gained currency as a viable method of estimating energy requirements in supply of goods/services to the economy.[8] It has been applied to study the dynamics of attaining self-sufficiency in rural systems with reference to India.[9] The emergence of second law-based analysis (Energy Analysis) is important in performing process analysis to effect conservation of energy.[10]

But these concepts and methods have not been adequately used in energy surveys in India. The lack of a common methodology, set of conventions and units makes comparisons futile and thus will be an obstacle in the generation of inputs in integrated energy-economics planning. But a more serious lapse is the lack of a commonly accepted set of calorific values, in particular of biomass fuels, and device efficiency factors. It would be helpful if a department connected with the issue of energy resources use in rural areas undertakes to gather experts to evolve detailed methodologies and conventions and standardize the relevant input values.

Energy in India: a flow-chart of energy use

We shall use the notion of flow of energy from a source to a use-sector in an analysis of the energy scene in India. Only the primary sources of energy will be counted as sources: electricity is treated as a secondary source. On the basis of statistics compiled from various sources[11], a flow-chart of energy use in India has been developed (Figure 1) for the year 1982/83. (This year was chosen, because reliable biomass data are not available after this period). The breakup of energy use by source and sector is given in Table 1.

The following inferences may be drawn from Table 1 and Figure 1:

1. Contribution of electricity to total useful energy in India is very small (3 percent); its contribution to domestic sector, the largest use-sector, is extremely small (less than 1 percent).
2. Biomass sources of energy contribute over 50 percent of total useful energy. This proportion has been somewhat stable since 70s, by which time it had declined from a value of 68 per cent in 1953-54.[12] Judging from this, it is reasonable to infer that biomass sources will remain the most significant source of energy in the decades to come.

But the economic planning does not reflect this realization. During the 7th Plan (1985-1990), the allocation for power sector was Rs. 320 billion (US $ 20 billion) out of a total allocation of approximately Rs.1600 billion (US $ 100 billion), whereas development of biomass received less than Rs.20 billion (about $ 1.3 billion). This is an outcome of following planning premised solely upon the economic value of energy conversion processes. This requires immediate correction in the years to come, keeping in view that development of biomass resources can be designed also to contribute to prevention of environmental degradation.

TABLE 1
A FLOW-CHART OF ENERGY USE (INDIA) 1982/83

	Units : Patajoules	Total	Electricity	Industry	Agriculture	Domestic + Others	Transport	Energy Sector own use
A	OIL	4737	343	880	231	905	2188	190
B	HYDEL	-411	411	-	-	-	-	-
C	NUCLEAR	27	27	-	-	-	-	-
D	LIGNITE	67	52	-	-	-	-	-
E	COAL	2593	944	1273	-	55	231	90
F	GAS	126	17	-	-	109	-	--
∝	TOTAL (A+B+C+D+E+F)	7961	1794	2153	231	1069	2419	295
G	VEG/AGRI WASTE	1407	-	100	-	1307	-	-
H	DUNG	770	-	-	-	770	-	-
J	FUELWOOD	4585	-	-	-	4585	-	-
β	TOTAL (G+H+J)	6762	-	100	-	6622	-	-
K	GROSS TOTAL (∝+ β)	14723	(470)	2253	231	7731	2419	295
L	ELECTRICITY	-	382	222	64	43	10	43
M	NET TOTAL	13309	-	2475	293	7774	2429	338
N	% NET	100%	-	18.6%	2.2%	58.4%	18.3%	2.5%
P	HUMAN FOOD ENERGY	2042						
			LOSSES CONV : 1324 T&D : 88					

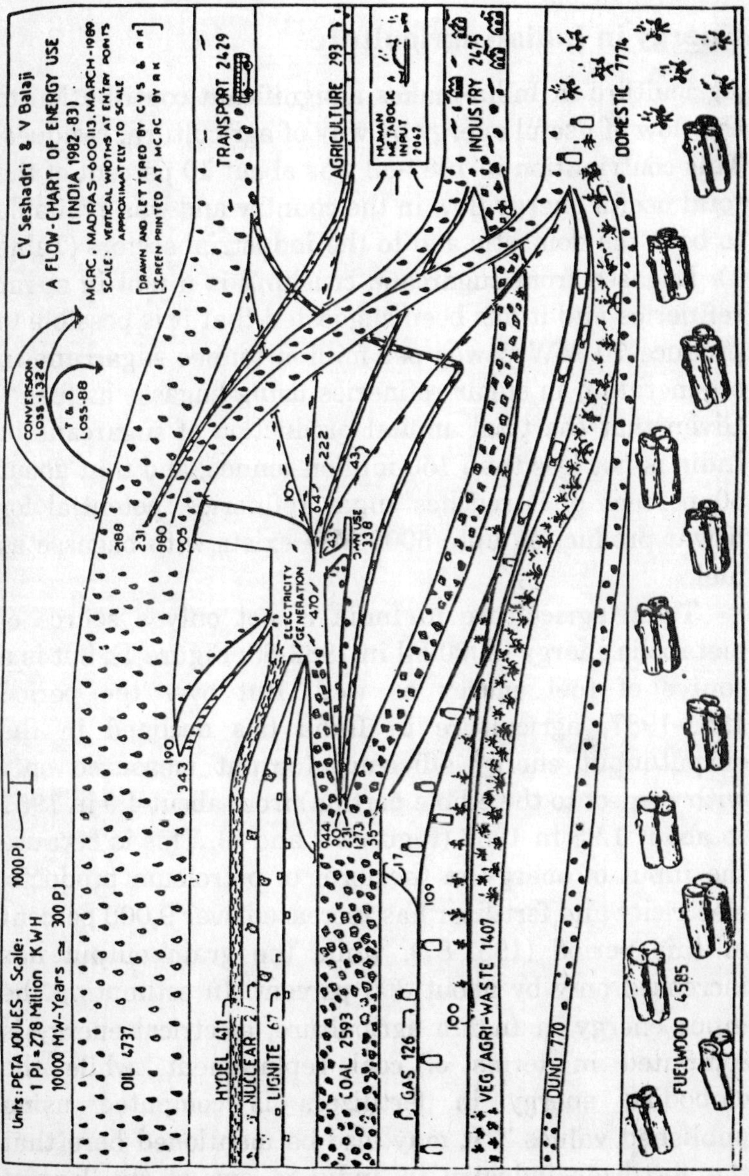

Figure 1 Flow-chart of energy use, India 1982/83.

Energy in Indian agriculture[13]

Agriculture in India makes a significant contribution to the flow of useful energy by way of agricultural residues. This contribution in 1982-83 was about 10 percent of the total useful energy flow in the country and was available to both the domestic and to the industrial sectors (Table 1). Bagasse from sugarcane crushing is a fuel in sugar refineries and it has been suggested that it is possible to produce 70 MW power per million tonnes sugarcane in cogeneration in sugar refineries using bagasse as fuel.[14] Given that the total annual production of sugarcane in India is not less than 150 million tonnes, and that about 50 percent of it reaches sugar refineries, potential for power production upto 5000 MW exists with bagasse as fuel.

Thus, agriculture in India is not only a source of metabolic energy (2000 PJ in 1982-83; Figure 1), but is a source of fuel energy as well. But over the period 1951-1987, agriculture in India has declined in the output/input energy efficiency (output measured only with respect to the edible portion) from about 4.0 in 1951 to about 1.25 in 1987 (Figures 2 and 3). This is because the input of energy in the form of petroleum products, electricity and fertilizer has increased over 9,000 percent in this period (1951-87), while the grain output has increased only by about 300 percent. In estimating the input energy in Indian agriculture, electrical energy is computed in terms of coal replacement, while the embodied energy in fertilizers is computed using published values.[15] It may also be mentioned here that the fertilizer industry in India is one of the largest "sinks" of useful energy, accounting for nearly 25 percent of the energy consumption in the industrial sector.

ENERGY CONSUMPTION IN INDIAN AGRICULTURE 1951 AND 1987

	Item	1951 Petajoules	1987 Petajoules	1987 Value / 1951 value x100
A	Electricity	3.3	468.0	14182 %
B	HSD LDO for Pumping	7.1	264.0	3734 %
C	Tractors	4.1	261.0	6397 %
D	Fertilisers	5.8	749.0	12914 %
E	Human Labour	70.0	147.0	210 %
F	Animal Labour	109.0	128.0	117 %
	Grand Total	199.3	2017.0	1012 %

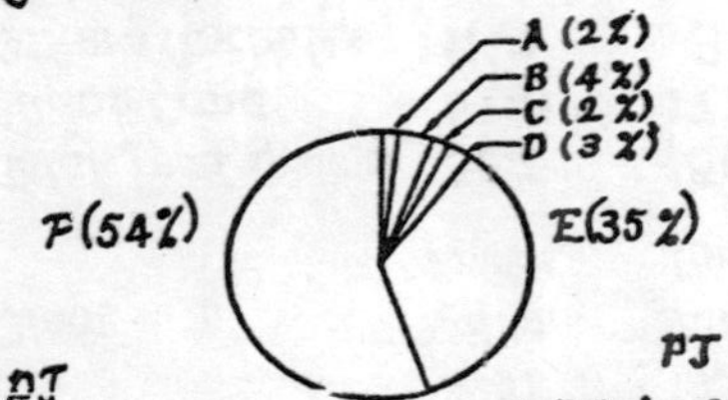

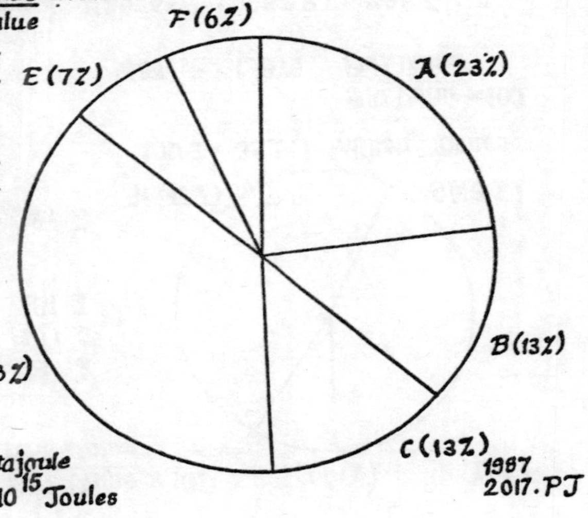

PJ = Petajoule

1 Petajoule = 1×10^{15} Joules

Figure 2 Energy consumption in Indian Agriculture 1951 and 1987

INDIAN CROP OUTPUT IN TERMS OF MASS & ENERGY
(ONLY EDIBLE PORTION COUNTED FOR ENERGY)

	Item	1951 Million tonnes	1987 Million tonnes	$\frac{\text{1987 Value}}{\text{1951 Value}} \times 100$
G	Total Food Grains	50.8	150.0	295 %
H	Sugar Cane	57.0	171.6	301 %
J	Others (includes fibre, spices etc)	11.1	31.1	281 %
	Total Mass	118.9	352.7	297 %

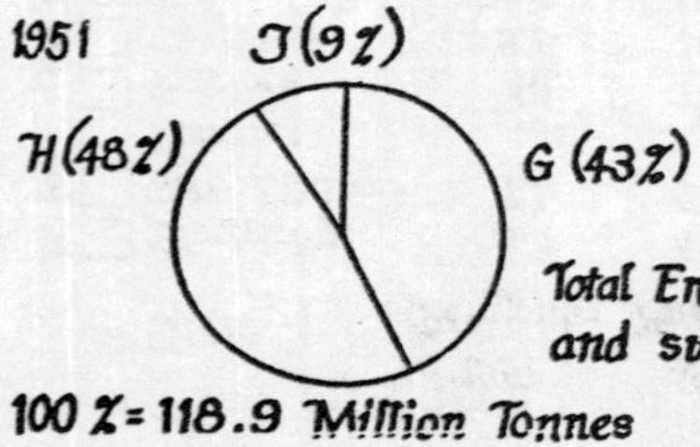

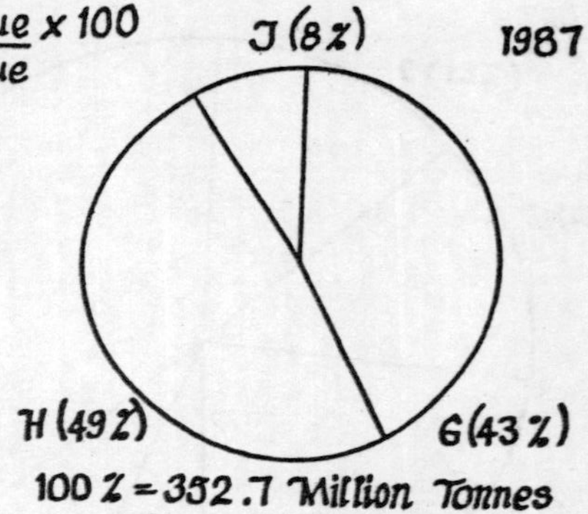

	1951	1987	$\frac{\text{1987 Value}}{\text{1951 Value}} \times 100$
Total Energy (foodgrain and sugar cane only)	826.6 PJ	2445.2 PJ	296 %

PJ = Petajoule
1 Petajoule = 1×10^{15} Joules

Figure 3 Indian crop output in terms of Mass & Energy

The increase in the energy requirement in Indian agriculture needs to be viewed with caution. If the present trend were to continue, the capacity of agriculture to contribute to energy flow in India might decline in the years to come.

Conservation through process integration

Conservation of energy has been advanced as a way of curbing energy demand, in addition to resorting to administration of prices. The latter approach in India has resulted in anomalous effects: it led only to augmented revenue and increased consumption of energy sources. Conservation has been taken up seriously in many developed countries, which have been able to bring down the rate of increase in the demand.[16] In India too, conservation is being pursued as a strategy of demand reduction in industrial as well as domestic sectors.[17] In the latter, use of improved *"Chulhas"* (cooking stoves) has been recommended as the most important means of reducing demand for fuel-wood in the domestic sector.[18]

The end-use oriented energy strategy (EOES) developed by Goldemberg and others has the advantage that energy saved in one sector can be used in yet another.[19] It has the further advantage of identification of the matching end use for a particular resource. Use of such an approach for conservation of biomass has been suggested in an energy plan for the Indian State of Karnataka.[14]

Another approach to energy conservation is to view it as a source of energy itself and integrate it with different uses. The concept of "process integration"[20] is based on the possibility of using low grade heat generated in one step of the process in another step where such heat would be useful; it has also been generalized to situations where different processes can be integrated along a chain of rejects. For example, it is possible to have coking-coal plant and thermal power plant operating on the same quantity of coal - the heat from combustion of coal is used in coking coal and subsequently in making steam.[21] Integration of biomass based processes offers one such possibility.

An example will be considered to illustrate this aspect. In a cane-molasses based distillery, the effluent, a high-pollutant, can be treated by anaerobic digestion to yield methane (minimum yield of methane would be 1 m^3/m^3 waste/day). The methane so generated can be used in the distillery, thus saving coal. However, if the effluent is distributed to nearby villages, where it is anaerobically digested in community plants to produce biogas, it may be useful in replacing fuel-wood. A hypothetical situation is given in Figure 4 and Table 2 where saving of 2,500 tonnes of fuelwood per annum is possible with effluent at the rate of 750 m^3/day from a distillery. This is to be preferred over saving of 600 tonnes of coal and 105 tonnes of HSD per annum.

The rationale here is that fuelwood for the cooking stove is a more immediate need than saving energy for the plant by reclamation of energy from effluent. Another factor to be considered is that for a C 4 plant like sugarcane, the CO_2 in the atmosphere may be fixed faster; hence a cycle that generates as much CO_2 as can be fixed locally can lead to alleviation of the Greenhouse Effect.

TABLE 2[22]
ENERGY YIELDS FROM SCHEMES IN FIGURE

I.	**Assumption and Norms**		
	Ethanol output from distillery (cane molasses-based)	:	10^6 litrrs/year
	Operating days	:	200/year
	Daily effluent output (Indian conditions)	:	750 m^3
	Methane requirement per average family (5 persons) per day (cooking only)	:	0.75 m^3
II.	If effluent were distributed (by trucks) to rural areas nearby distillery	:	
	No. of daily trips (10 m^3 effluent per trip)	:	75
	Distance per trip (average)	:	50 km
	HSD required per trip	:	10l
	Annual requirement of	:	150,000 l
	HSD	:	=105 tonnes
	Methane available per day	:	750 m^3
	No. of beneficiary families	:	750/0.75=1000
Savings:			
	Biomass requirement for 1000 average rural families over a year	:	3500 tonnes
	of this, Fuelwood requirement	:	2500 tonnes
II.	**If the methane generated is used within the distillery**		
	Annual production of methane	:	0.15 x 10^6 m^3
	Coal replacement (1 million m^3 methane = 4000 tonnes coal replacement)	:	600 tonnes
III. a.	**Total Fuel Expenditure (Annual)**		
	Coal: 600 tonnes : HSD:105 tonnes		
b.	**Total savings**		
	i. fuelwood	:	2500 tonnes;
	ii.dung and residues	:	1000 tonnes

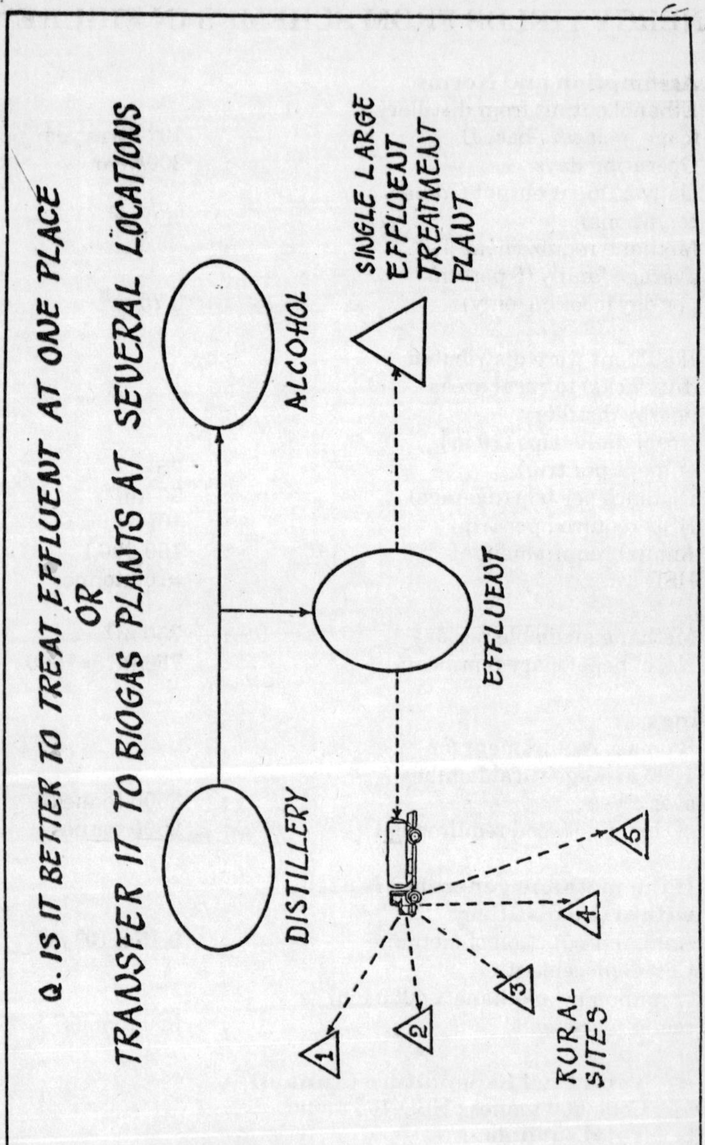

Figure 4 Energy conservation through Microbial Cycles - Process Integration.

Conclusions

1. There is a need for Third World countries to decide on a uniform methodology for inclusion of "commercial" as well as "non-commercial sources" of energy in their plan strategies.
2. There is a need to recognize that for several more years to come domestic energy demands will be the major demand and this will be met substantially only from biomass sources.
3. District level energy surveys even in biomass-rich districts seem to lead to the conclusion that expenditure on new biomass plantations is not proportionate to its harvest.
4. Process integration using microbial cycles offers the best hope in newly industrializing countries. The use of carbohydrate waste in tropical countries for domestic cooking and heating seems to be the most fruitful pathway.

References

1. Planning Commission *Seventh Five-Year Plan: A mid-term appraisal,* New Delhi, 1987.
2. P.P. Christensen, "Historical Roots for Ecological Economics - Biophysical Versus Allocative Approaches", *Ecological Economics, 1,* 1989, 17-36.
3. H.J. Barnett, and C. Morse, *Scarcity and Growth: The Economics of Natural Resource Availability,* Baltimore: Johns Hopkins University Press, 1963.
4. R.M. Solow, "The Economics of Resources or the Resources of Economics" *American Economic Review, 642* , 1974, 1-14.

5. Georgescu-Roegen, *The Entropy Law and the Economic Process* Cambridge, USA: Harverd University Press, 1971.
6. G.R. Tyner, R. Costanza and R.G. Fowler "The Net Energy Yield of Nuclear Power", *Energy*, 13, 1988, 73-81.
7. H.T. Odum, "Self-organization, Transformity, and Information", *Science 242,* 1988, 1132-1139.
8. IFIAS (International Federation for the Institutes of Advanced Study), *Energy Analysis,* **Workshop Report No.6,** M.Slesser, (ed), Solna, Sweden, 1974.
9. M. Slesser, C.W. Lewis, I. Hounam, C.V. Seshadri, R.N. Roy, N. Jeejibai, N. Keshavaraj, R. Manoharan, G. Raja, V. Subramani, S. Thomas and G. Venkataramani "Biomass Assessment in Third World Villages via a Systems Methodology", *Biomass, 2*, 1982, 57-74.
10. G.Wall, *Energy- A Useful Concept*, Goteborg, Sweden: Chalmers University of Technology, 1986.
11. ABE (Advisory Board on Energy), *A Perspective on Demand for Energy in India upto 2004-05,* New Delhi: Advisory Board on Energy, Government of India, 1984. D.R. Pendse, *Statistical Outline of India 1987-88*, Bombay: Tata Services, 1988. T.R.Satish Chandran, "National Energy Balances of India" paper presented at the 12th congress of the *World Energy Conference,* New Delhi, 1983. *Statistical Abstract*, Central Statistical Organization, Government of India, 1986.
12. P.D. Henderson, "Energy Resources, Consumption, and Supply in India", in *Energy in the Developing World: The Real Energy Crisis*, V.Smil and W.E. Knowland, (eds) New York: Oxford University Press, 1980, 172-193.

13. C.V.Seshadri, *"Energy in Indian Agriculture"*, **Review of Agriculture, The Hindu**, Madras, 1989.
14. A.K.N.Reddy, *A Development-Focussed End-Use Oriented Energy Plan for Karnataka* (**Chapter V of Karnataka: Perspective Plan 2001**), 1988.
15. M.Slesser, and C.Lewis, *Biological Energy Sources*, London: E & F N Spon, 1979.
16. **UNEP (United National Environment Program)**, *Energy Conservation in Developing Countries*, Nairobi : UNEP, 1986.
17. ABE, *Energy Conservation: Challenges and Opportunities*, New Delhi: Advisory Board on Energy, Government of India, 1986.
18. R.K. Pachauri, and L. Joshi (eds), *Energy Policy Issues*, 4, New Delhi: Tata Energy Research Institute, 1989.
19. J. Goldemberg, T.B. Johansson, A.K.N. Reddy and R.H. Williams *Energy for a Sustainable World*, New Delhi: Wiley-Eastern, 1988.
20. C.A. Berg, "Process integration and the second law of thermodynamics: future possibilities", *Energy 5,*1980, 733-741.
20. E.P. Velikhov, V.M. Mastennikov, A.E. Shendlin, E.E. Shpilrain and V.Ya. Shtrenberg "Integrated and industrial processing as a method of energy conservation" *Energy, 12,* 1987, 1083-1095.
21. T. Vatsala, Augmentation of Biogas from Distillery Effluent using C. Fresindi, Project Report submitted to the Department of Non-Conventional Energy Sources, Govt. of India (from MCRC, Madras).

Towards Alternative Paradigms of Development

V. Sudarsen

I

Modernization and Development though different semantically, have come to acquire the same substantive meaning in terms of their conceptual usage. Conceptually, the taxonomical gradations used by the theorists of Modernization and Development overlap to a large extent. Dube, while discussing the concept of Modernization and Development, identifies three characteristics common to both of them:[1]

"Modernization and Development have been built on considerable exploitation of certain segments of the society and have involved a degree of ruthlessness. Imperialism aided them substantially. Their astonishing accomplishments have caused and are still causing considerable social injuries ..... Attempted justifications of the inevitability of this injury-causing dimension of the process were made in the initial phases .... Now it is being openly challenged.

Second, modernization and development are multilinear/multipath processes ... alternative paths can be followed.

Third, neither can be visualized as continuous and unending."(p.5)

Use of Science and Technology through a structurally organized means is stressed in the process of modernization as propounded by the western social scientists. But the reality is that a minority, the advantaged sections of the society, has greater access to both and this minority manipulates the organizational structure in such a way as to derive maximum benefits, while the vast majority languish in neglect. To quote Dube (*ibid*) again : "...as no social indicators are built into the modernization formulation, its social objectives are neither specific nor quantifiable. The dimension remains blurred...It should be noted that the objectives of modernization have not been related to the paramount consideration of social justice" (p.27).

In fact, the Modernization and Development Paradigm remains obscure on several counts. It remains silent on the vital issue of the poverty of masses. The two basic questions of Modernization and Development 'for whom' and 'for what' are not answered at all.

The protagonists of 'structural model' precisely try to answer the questions that remained unanswered by the propounders of Modernization and Developmental paradigm. The former clearly analyzed the major issues in Developmental theory and conclude that: underdevelopment is not the original state in the evolutionary process; development does not necessarily travel from centre to periphery, while in fact development and underdevelopment are two sides of the same coin; and a dualistic situation at the international and national levels is a part of capitalist development.

With this background one can only visualize the future of the Developing Countries in pursuing a course of development which is need-based, endogenous, self-reliant, ecologically sound and above all based on structural transformation.

II

In this paper, I have taken up a few issues related to the question of development in the Indian context. I have also provided illustrations, wherever possible, to make my point of view clear. The issues I have dealt with here are basically the forest and land-use rights vis-a-vis ecology in the overall context of development in India.

Forest Policy: a distinct continuity

As a consequence of Industrial Revolution, England became a major manufacturer and exporter of the world. A new industrial capitalist class developed which gained a predominant position in British economy and politics. The colonial economic policies were virtually a reflection of the needs of the industrial capitalist class. As a result, India was compelled to export raw materials to pay her arrears of interest and dividend.[2] It is not surprising to find that the earliest plan and execution of rail transport links, to start with between Bombay and Thane, were done with a view to providing greater control over resources in colonies which was crucial for the expansion of British capitalism.[3]

With the increasing timber needs of the British, the first Forest Policy was formulated by Lord Dalhousie in 1855 with the ultimate objective of providing the state the reigns of control of the forests in India. The Imperial

Forest Department was created as a part of this policy and with the avowed claim of 'Scientific Management of Forest Resources'. In the year 1865, the first *Forest Act* was passed which empowered the Government to declare any land as Government Forest Land, further, empowering the state governments to prescribe punishments for the breach of the provisions.[4]. It is quite interesting to note here that Britain was almost completely denuded of forests before the British started the 'Scientific Management' of forests in India.

After the 1865 Forest Act, there were no less than ten Central Forest Acts (or amendments to the existing Acts) and many Acts relevant to forest management, such as the Cattle Trespass Act (1871), the Wild Life Preservation Act (1972). etc.[5]

One of the major concerns of all these legislative measures was to keep the forest cover stable. Our National Forest Policy (1952) also lays down that 33 percent of the total area of the country should be under cover, with 60 percent cover in hilly regions. In no state or Union Territory of India do we find the forest cover to the extent as laid down in our Policy almost four decades ago, except in Manipur (Table 1).

Table 1
Change in the Proportion of Forest Cover in Different States /Union Territories (UTs) during 1972-75 and 1980-82

State/UT	% of Forest area in relation total geographic area in the State/UT		%change in forest cover in terms of forest area in 1972-75
	1972-75	1980-82	
Andhra Pradesh	17.72	14.61	(-) 17.56

State/UT	% of Forest area in relation total geographic area in the State/UT		%change in forest cover in terms of forest area in 1972-75
	1972-75	1980-82	
Assam	26.81	25.21	(-) 5.98
Bihar	13.06	11.58	(-) 11.23
Gujarat	4.83	2.58	(-) 46.54
Haryana	1.71	0.91	(-) 47.03
Himachal Pradesh	27.08	15.40	(-) 39.45
Jammu & Kashmir	10.05	6.46	(-) 35.69
Karnataka	15.37	13.38	(-) 12.97
Kerala	22.15	18.98	(-) 14.34
Madhya Pradesh	24.52	20.37	(-) 16.90
Maharashtra	13.22	9.86	(-) 25.40
Manipur	67.49	60.70	(-) 10.06
Dadra & Nagar Haveli	36.12	25.51	(-) 60.45
Meghalaya	63.98	55.39	(-) 13.43
Nagaland	49.33	48.97	(-) 0.72
Tripura	60.40	49.03	(-) 18.83
Orissa	31.06	25.31	(-) 18.51
Punjab & Chandigarh	2.22	0.99	(-) 55.45
Rajasthan	3.30	1.75	(-) 47.12
Tamil Nadu	12.83	10.14	(-) 20.32
Uttar Pradesh	8.79	7.14	(-) 18.74
West Bengal	9.50	7.38	(-) 22.33
Sikkim	24.12	39.49	(+) 63.17
Arunachal Pradesh	61.54	62.34	(+) 1.29
Delhi	1.21	0.67	(-) 44.4
Goa, Daman & Diu	32.04	29.90	(-) 6.2
Mizoram	68.72	56.76	(-) 13.63
Total variation (%) from 1972-75 to 1980-82			(-)17.18%

Source: National Remote Sensing Agency, 1984.

One can perhaps make a generalization in this context, i.e., there was very little change in the pattern of utility or the rights of the people over the forests in the periods of British imperialistic rule and Indian Independent governance. It is not my intention here to delve deep into the question of continuity in policy implications but only to point out such a distinct continuity.

The depleting forest cover is a broad reflection of the Anthropogenic interference. The Anthropogenic Acts include our forest resource needs for industry and also as a by-product of other developmental activities. The crucial question that needs to be asked is how to draw that 'golden mean' between meeting the human needs and maintaining forest cover. How is this determined? Scientifically or arbitrarily?

Talking about the environmental issues in Kumaon area, Singh says:[6]

> "Reduced forest cover accelerates soil erosion, increasing silting of waterbodies, drying up of springs, the replacement and disappearance of species, and increased ratio of energy expended in fodder and fuel collection to energy expended in direct agricultural activity and consequent high level of drudgery of women folk, are but some of the tell-tale symptoms of environmental ill health..... It is advisable that management of such ecological resources upon which local communities depend should be decentralized and *these communities should be given an effective say over the use of these resources* (emphasis mine)."

The above observation brings in the complexity of environmental problems involved in unplanned exploitation of forests. In this context, I deem it appropriate to quote from Agarwal's lecture on "Ecological destruction and emerging patterns of poverty and peoples protests in rural India" (As quoted in Joshi, 1990).[7] Agarwal says that "The pattern of global exploitation that we see at the global scale simply reproduces itself at the national level. Exactly what the western industry does to the Third World environment, the Indian industry does to the Indian environment".

I will illustrate the point made by Agarwal with an example from Andamans. Andamans is the habitat of rare evergreen rain forest. In the Little Andaman Island, there was tremendous deforestation as a consequence of government's policy to fell the trees for auctioning the timber. As an afforestation measure, the Andaman forest department decided in early eighties to go for mono-species afforestation program. Firstly, a single species is no substitute for multispecies forests that are being replaced. Secondly, another interesting and vital point came to the fore. It was red oil palm which was used as the single species and it was known that one giant edible oil producing company of India influenced the decision considerably. The reasons are not far to seek - the oil prices were high and India was importing palm oil from Malaysia at that time. The result was that, all the considerations of 'Scientific Management' were swept under the carpet.

In the name of modernization and development in the context of forestry management, what is being done is the steady transformation, in a particular direction, of the very basis of nature. From a biological point of view, it is a reduction from the set of multispecies diversity to a series of high yielding mono cultures. At the social level, the transformation is in terms of forest being a supporter of household and community needs to the one which is mainly meant to meet industrial needs and the urban rich (Agarwal as quoted in Joshi 1990).[7]

Biologically, the greatest amount of diversity is seen in the tropical and subtropical regions. Vasilov (1951) traces the origin and variations among different centres located in the developing countries.[8] These are the fountain heads of genetic variation. Most of the high yielding crop varieties developed by the developed countries depended for the germ plasm from these centres of diversity. Khoshoo[9] rightly says : "Concern has been expressed with regard to the patenting of varieties and the role multinationals will play. The fear is that more and more material will be in the hands of less and less number of people and we who are in the centre of diversity may be held to ransom. On account of the prospect of application of genetic engineering all the information generated will be confined to studies and books of the companies. Scientists are under contract not to publish or even talk about it." (p.95).

The apprehension expressed is no more a matter of theoretical possibility. There are definite attempts to control the development of all new life varieties through genetic engineering research for private profit. The recent GATT negotiations in Uruguay witnessed some of the developed countries demanding adoption of national laws towards private ownership of life forms as a part of intellectual property rights. What in essence they are demanding is that there should be private ownership over life itself. "There is an urgent need for concerted action in several areas by the citizens, national governments, non-governmental organizations..." and there should be an universal agreement that *"life forms are the common heritage of mankind"* (emphasis mine).[10] There should also be a concerted effort to solve the problems of common man through the genetic engineering research, like agriculture, forestry, animal husbandry and health.

III

This brings me to the second aspect of the question of development. Technological achievements like multipurpose dams, thermal and nuclear power stations were certainly landmarks in the S&T history of independent India. But these achievements cannot be viewed in isolation, independent of the human and other costs involved. It is not the technological capability *per se* that is being questioned here, but the lack of a holistic dimensions of technology which need to be taken into account, while discussing the developmental perspective and a total rethinking required in this regard.

In the past one century, about three thousand major and medium dams have been constructed in India in the areas of dense natural forests inhabited by mostly tribals. According to the Planning Commission's report in 1979, there were 1551 large dams still under construction in various states. According to the estimates of the working group on the development of scheduled caste during the Seventh Plan period, every development project invariably displaced large proportions of tribals ranging from a minimum 52 percent to 100 percet from one project to the other. It is not only construction of the dams but various other projects also caused such hardships. The National Testing Range at Balipal is a case in point. The original number of persons to be relocated was estimated to be one lakh persons coming from 100 villages. When the people organized a stiff resistance, somewhere along the government came down to a number of 60,000 evacuees from 68 villages. "It is again an evidence of the same insolent non-challance towards the cheap human commodity".[11]

Similarly, the Narmada project if fully implemented would submerge about one lakh hectares of land, including rich forest, and would uproot about 150,000 people inhabiting 190 villages. According to the estimates of the Department of Environment and Forest, the timber loss alone from the submerged forests will be of the tune of Rs. 40,000 crores.

Apart from the questions purely environmental in nature, one must ask "why have the weaker sections of Indian people invariably been the victims of large public projects? Or conversely, why have lakhs and lakhs of displaced or ousted people never been rehabilitated properly?[12] To put it in a nutshell, most of the tribals in India are a dispensable commodity for development projects, "having no occupancy rights to land or habitat, they are not *ipso facto* entitled to compensation ....." *(ibid)*. Just to illustrate this point, one can glean through the Government of India's policy of land acquisition for National Testing Range at Balipal "Out of the 21,000 acres required for the Range, 11,000 acres were considered 'encroached' by local farmers and no compensation was paid for that land".[13] During the recent *Narmada Tribunal Award* concerning Narmada dam project, the Gujarat government vehemently argued that the displaced tribals have no property rights. *It also claimed that whatever compensation in terms of homestead land is being granted to the displaced Tribals is an act of "benevolence."*

These experiences have to be assessed by keeping in mind the fact that India ratified the ILO (107) *CONVENTION* regarding the international laws applicable to the forest dwellers and also was a party to *The Human Rights Convention, The International Covenant of Civil and Political Rights, The Convention of Aboriginal Rights,* which recognizes the laws not only in terms of human dignity but also in terms of laws of culture and community resource base.

The specific instances and experiences which I tried to highlight form only a miniscule part of the complex problem. But from this brief discussion it becomes clear that technological progress is not axiomatic with the development process but they have to be provided proper linkages.

The concept of sustainable development is gaining ground all over the world. The realization of this concept in practice requires the full support of the scientists and policy planners on one hand and the political will of the ruling government on the other. When I mention this, I am fully aware of the linkages between the social system and the political institution. The nexus between the groups with vested interests bent upon exploiting the natural resources at any cost and the political power bases can be countered only by people's participation. If a majority of people of India opt for a better living for all "with few risks" than higher standard of living for a few "with greater risks", it only means that there is a need to have a closer look at our developmental priorities.[14]

Development with the parameters straight-jacketed into a linear mould is what is being used to-day. The existing paradigm of development needs to be assessed afresh and alternative paradigm developed in the light of such an assessment. Unless efforts are made to evolve alternative paradigms of development, we are sure to end up more in the quagmire of ecological, economic and social problems instead of solving them.

References

1. S.C.Dube, *Modernization and Development : The Search for Alternative Paradigms,* Delhi : Vistaar Publictions, 1988.
2. Bipin Chandra,"The Impact of British Rule" in K.P.Rungachary (ed), *Freedom Struggle,* Delhi: National Book Trust, 1972.
3. I.Walllerstein, *The Modern World System,* New York, 1974.
4. P.Prabhu, "Pre-Independence Forest Legislation and the Tribals, *Legal Perspectives* (Mimeo), LRSA-Document File, 14.
5. National Commission on Agriculture, *Report on Forestry,* Vols. 8 & 9, Delhi: Ministry of Agriculture & Irrigation, 1976.
6. J.S.Singh, "The Environmental Problem: Issues and Challenge" in K.S.Valdiya (ed) *Kumaon: Land and People,* 1988.
7. P.C.Joshi "Uttarkhand and Indian Renaissance" *Mainstream,* Vol.XXVIII, No.17, 1990, pp.17-23.
8. V.I.Vavilov, "The Origin, Variation, Immunity and Breeding of Cultivated plants", *Chronica Botanica,* Vol.13 1951, pp 1-364.
9. T.N.Khoshoo, *Environmental Priorities in India and Sustainable Development,* Presidential Address to the 73rd Session of Indian Science Congress Association, Delhi, 1986.
10. S.J.Patel, "Gene Revolution and the South", *Mainstream,* Vol. XXVIII, No.17, 1990, pp 9 and 28.
11. N.S.Reddy, "The Key Issue: Development Displacement and Resettlement" in V.Sudarsen, and M.A Kalam (eds). *The Uprooted people,* Delhi: Gian Publishers, 1990.

12. Chatrapathi Singh, n.d."Rehabilitation and the Right to Property" *Legal Perspectives* (Mimeo), LARSA-Documentation File No.14.
13. V.Sudarsen, and M.A Kalam, "Introduction" in V.Sudarsen, and M.A.Kalam (eds) *The Uprooted*, Delhi: Gyan Publishers, 1990.
14. *The Hindu*, "Far From Apathetic" (editorial), May 18, 1989.

Science and Technology in Society : Diversity - The Central Theme

Ashok Jain

The metaphor 'Science in Society (rather than 'Science and Society') belongs to an emerging paradigm in natural and social sciences in which non-linearity, diversity and asymmetry occupy a central place. This metaphor enables us to launch a new project for understanding science, technology and society in a holistic manner. It signals a movement away from a reductionistic and mechanical view of the world and celebrates the exhaustion of Newtonian mind-set that considered asymmetry and diversity as aberrations and aspired to recreate reality as ordered fragments fitting into a giant clockwork.

The response of this conventional mind-set rooted in mechanistic view of the world to rich diversity inherent in social process has been the slogan 'two cultures'. One, the culture of science with its preoccupation with order and symmetry prospering in synergetic alliance with technology of mass production and the other concerned with understanding the tapestry of diverse social fabrics.

The slogan enabled the culture of science to pursue it's goal of spreading the message of uniformity, diversity and heterogeneity being the pursuits of social sciences. The technology arm of science, within this mind-set, glorified uniformity by propagating the virtues of standardisation and efficiencies of large scale centralised production systems.

For the adherents to the mechanistic view of the world, the slogan of 'two cultures' thus provided a comfortable thought space as well as sufficient opportunities for capturing resources through industrial technologies. Increasing concerns for environmental and other impacts of industrialisation required the culture of science to articulate it's stand on that part of reality that it had conveniently left to social sciences. The response was the metaphor 'Science and Society'; essentially an expression of willingness to enter into a dialogue with social sciences. A dialogue, the agenda for which is set within the framework of 'two cultures'. A dialogue in which science participates as an exponent of mechanistic view.

The metaphor 'Science in Society', on the other hand, signifies a willingness to move out of the mechanistic view of the world. It inherently rejects the premises of 'two cultures' as it has no stake in promoting uniformity and homogeneity. By cutting lose from the entrenched reductionistic world view, it at once comes under the fold of a new paradigm in which non-linearity and diversity occupy a central place. A new paradigm that does not require looking at reality through two cultural polaroids, one that of science to see order and homogeneity and the other of social science for disorder and heterogeneity. It is in this context that the metaphor 'Science in Society' gives us an agenda that is much deeper than a simple dialogue between natural sciences and social sciences.

A brief elaboration of this emerging paradigm in science and technology follows.

Emerging science and technology paradigm

Henri Bergson, one of the foremost philosophers of this century observed, 'the universe is not made, but is being made continually. It is growing, perhaps indefinitely ....' The amalgamation of regularity and irregularity is the very essence of growth of the physical or material world.

To understand the material world, the central concern of natural sciences is shifting from uniformity and order to seek answers to how the universe is advancing from a featureless simplicity to complex forms on all scales from molecules to galactic clusters. A characteristic feature of complexity, chaos and disorder is that, by very nature, complex forms have a high degree of individuality. For example, we recognise a tree as a *mango* tree but no two mango trees are the same. This individuality and diversity cannot be comprehended by the conventional scientific approaches rooted in the Newtonian paradigm. The conventional paradigm in which change is smooth and continuous is unsuited to deal with irregular real systems because:

- complexity often appears abruptly and not through slow smooth changes;
- complex systems often have large number of degrees of freedom;
- systems are open to interaction with complex environment that drives them; and
- these systems are mainly non-linear.

An entirely new science paradigm is emerging to replace the three hundred years of entrenched philosophy; natural sciences are moving away from concerns with uniformity to concerns with diversity and complexity (Davies, 1989). A similar movement is seen in technology.

The direction of modernisation of technology is towards meeting the diverse requirements of individuals; from 'mass production' of uniform goods and services to those tailored to the needs and likings of individuals.

> "Computer-integrated manufacturing (CIM) or 'flexible automation' is a revolutionary response to the 'complexity- reliability-variety' barrier. In brief products are becoming more complex, quality control is becoming more difficult and customers are less satisfied with standardisation. Managers of manufacturing firms face several critical questions and some strategic choices. Among them are the following: How must the firm change internally to compete successfully in the future? What technologies must be mastered and adopted 'in house', and what can be safely contracted out? Is there an important distinction to be made between 'human-centred' and 'machine- centred' approaches? (Ayres, 1992)

Not only the final technological products, but the *manufacturing process* itself is undergoing a change. It is being realised that just as natural phenomenon is to be understood in terms of open systems interacting with complex environment, the manufacturing process is also essentially an interconnected complex system, in which one step or stage depends upon all the others i.e., a holistic approach is necessary. The mechanistic view of nature (ordered fragments fitting into a clockwork) had influenced manufacturing technology resulting in

Taylorism that claimed that the performance of an enterprise could be maximised by maximising the performance of individual task elements. That approach has come into serious questioning.

The paradigm shift in 'manufacturing process' component of technology has come about in several stages (Ayres, 1992):

1880s : **The English System** : Application of power-driven machines to the old production methods used in the pre-industrial period.

1850-1920 : **The American System** : This emphasised standardisation of products and interchangeability of parts.

1920-1960 : **Taylorism** : This emphasised mechanical integration, assembly line, vertical integration and optimization of individual tasks.

1960-1985 : **Quality Control** : This added to Taylorism, new techniques such as statistical quality control, total quality control and applications of computers to individual functions.

1990- : **Flexible and Holistic Approach** : The emphasis is on functional integration, flexibility, decentralisation, horizontal management, team work and networking.

This new emerging paradigm of technology, moving away from ordered and standardised products and processes to complex, flexible and holistic view, in its essence, is analogous to the emerging paradigm in natural sciences.

It is under this new overall paradigm of science and technology that the concept of human-centred systems has emerged in Europe (Gill, 1990) and in Japan (Sato, 1991). According to Sato, in Europe this concept is rooted in the cultural diversity amongst the European nations.

In summary the concept of human-centred systems can be defined as (Cooley, 1987) 'human centred systems ought to support human skills and acquisition of new skill and the design of these systems should consider technology as a tool and not a machine'.

The implication of this definition is that a 'machine' (or production process and technology) must be situated in the context of creativity, imagination, intuition, expertise etc. — the characteristics of 'skills' of an individual.

The above discussion is to emphasise that science and technology no longer consider complexity and diversity as hindrances that are to be avoided but these are the very core on which our understanding of nature and of production systems is being built (Hansson, 1991). This is a radical change from the conventional mind-set and makes the slogan of 'two cultures' and with it the metaphor 'Science and Society', redundant.

The paradigm within which the new metaphor 'Science in Society' is placed characterises a widespread global change. 'Globalisation of trade, finance and production systems coupled with emphasis on decentralisation and liberalisation are leading to hitherto unknown interconnections, between different social systems internationally. The complexibility, plurality and instabilities inherent in these interconnections that have led to the emergence of a new social, economic and technological paradigm, (Xinhua, 1992).

It is interesting to note that discussions on a new developmental path, generally, referred to as sustainable development are also connected with the emergence of this new paradigm. We now proceed to show this interconnection.

Environmental concerns — the crisis of a mind-set

It is well understood that the choice before the world community is not between development and environment but between environment-sensitive and environment-insensitive development. The basic elements of environmentally sensitive form of development are (Sachs, 1992):

- Social sustainability : The aim is to build a civilisation of beings with equity in *assets and income distribution*.
- Economic sustainability : Economic efficiency should be evaluated in *macroscopic terms* rather than only through the criterion of micro-entrepreneurial profitability.
- Ecological sustainability : Intensifying the use of the resource potential of *diverse* ecosystem with minimum damage to the life supporting systems.
- Spatial sustainability : Balanced distribution between rural-urban through exploiting the potential of *decentralised* industrialisation linked with new technology.
- Cultural sustainability : Devising *local* ecosystem-specific, *culture-specific* and *site-specific* solutions.

Clearly environment-sensitive development requires strategies to deal with diversities in income, eco-systems, cultures and a movement towards decentralisation. The conventional paradigm of science and technology considered diversity and complexity as hindrances and aberrations. Science and technology proceeded by passing on these dimensions of reality to social sciences on one hand and by trying to eliminate these through usual technological mechanisation and standardisation on the other hand. The inherent contradiction and limitations of this approach indicated a dead end, a saturation or plateuing of progress. An exhausted science and technology paradigm had to declare a crisis situation — the environmental crisis. Should one still look for ways and means of overcoming the crisis within a paradigm that itself has led us to declare the crisis? Should one still stick to the metaphor 'Science and Society'? Perhaps a little more probing into the relationship between sustainable development and the new paradigm is required.

Flexibility, diversity and sustainable development

Let us consider only one facet of development viz., industrial production. In its simplest form, a production technology is a method of combining the following five Ms:

M1 - Machine/know-how

M2 - Materials, including energy

M3 - Money

M4 - Management and manpower

M5 - Market and user needs

Any production process requires a synchronisation of the five Ms. Artisanal technologies use simple tools for M1, locally available materials (M2), minimum investments (M3), family unit for management and manpower (M4) and small surrounding markets (M5). Available industrial technologies are based on large M1, M2, M3 and M4 procured from different locations to produce standard goods M5.

It is clear that machine or know-how i.e., M1 in itself does not specify the technology. Indeed it is M2, M3 and M4 that decide the type of machinery or know-how that can produce M5 i.e., meet the user or market demand.

The main limitation of conventional technologies in use is that, in the absence of detailed knowledge of materials (M2) and of know-how (M1), only a limited type of combinations of Ms i.e., technologies are tried.

Developments in informatics, biology and material sciences have improved our understanding of process know-how i.e., M1 and of materials (M2) many-folds. Microprocessors enable us to make machinery that is flexible. Modern biology gives us a variety of new bio-chemical processes. This knowledge, can now be used for combining Ms in new ways. Thus, new combinations or new technologies can be developed to give production processes that have better environmental features.

A new small-scale industrialisation strategy as an environmentally-sensitive alternative economic and institutional mechanism is thus emerging (James et al., 1991). For example, a flexible production and batch processing system has been proposed by the Centre for Technology and Development (1991) for modernising leather, fruit processing, pottery and ceramic, metalware and other such artisanal and conventional production units. Flexibility and descaling being done through micro- processor control systems.

The technology being developed has the following features:

- Low cost automation tools that can be handled by relatively unsophisticated plant operators (e.g., family members of vegetable/fruit producers).
- Instead of a fixed plant for a specific product, the technology uses a generic plant. Several products

can be processed. For example, different types of apples, oranges, tomatoes etc., can be converted into fruit juice of standard quality.

- Descaling is possible so that individual vegetable/fruit grower or small cooperatives can run production units, small volumes being compensated by the range of products that can be processed.
- Reduction in wastage. Special sensors combined with control systems will enable use of fruits and vegetables of different qualities which otherwise go waste.

Environmental advantages are obvious. Wastage is reduced, transport of material and associated energy consumption and pollution is drastically cut. The plant itself is least power consuming as it uses microprocessors for quality control of process parameters and human labour for cutting and crushing of fruits/vegetables. The capital/labour ratio is better than conventional industrial technologies. By locating small production units closer to the producers of raw materials migration and associated problems of urbanisation are avoided. Similar scenario emerges for pottery and ceramics, leather, metal ware etc.

The above example illustrates that the concept of modernisation of industry is a movement away from large scale centralised production system to decentralised flexible production systems that have better environmental features. This concept of 'modern' industry is opposite to the conventional notion in which large centralised production units with high efficiencies were considered modern although they had very low or negligible flexibility. We, therefore, find that the concept of 'modern' industry is in tune with the new paradigm.

Having discussed the new paradigm and it's relationship with environment-sensitive development, we revert back to the question of a project under the agenda 'Science in Society'. Are there any methodologies on the horizon for pursuing this agenda?

A possible approach for studying 'science in society'

The changing scenario may be summarised as:

- Science paradigm is shifting towards an understanding of complexity and diversity. Non-linear, open and self-organising systems are attracting attention.
- Technology paradigm is also moving towards complexity, flexibility, decentralisation and human-centredness.
- Flexible manufacturing coupled with bio-technology and the science of new materials has already given techniques for decentralised productions with better environmental features.
- Political and economic forces are also leading to liberalisation (i.e., human-centredness) and global connectivities.

As a consequence of the above changes, science and technology concerns now cover not only research and development but a much broader issue and that is of innovation in society (Solomon, 1990). Recognising that innovation grows through heterogeneous processes of social and technical change, the concept of Techno-Economic Networks (TENs) has emerged for understanding innovation (Callon et al, 1989).

TENs are not networks of the type one is familiar with in telecommunications or railways. Nor are they networks of families of actors. TENs are a mix of humans and non-humans with identities of different sorts. The dynamic behaviour of TENs gives rise to innovation and it's diffusion and can be studied through translation operations that identify the relevant prescriptions. This appears to be a promising methodology for pursing a research agenda under the metaphor 'Science in Society'.

The emerging world view places diversity at it's centre. All the disciplines of natural and social sciences appear to move towards a paradigm that does not treat heterogeneity as an aberration, rather it is viewed as the very source of innovation and future movement. This is a major departure from the conventional paradigm.

The conventional paradigm has well served humankind for the last three hundred years. The industrial revolution in Europe triggering off it's exploitation for the benefit of man. As is inherent in any evolutionary growth, a particular path reaches saturation and then growth occurs along a new path. Environmental crisis is a sign of saturation of the conventional paradigm. The emerging paradigm with diversity as it's central theme may be the new path for humankind. Developments in science and technology are beginning to emerge along the new path. Tools and techniques for examining strategies for moving along the new path are also available. The metaphor 'Science in Society', it is hoped will help create a collective recognition of this change.

Reference:

1. R.U. Ayres, *CIM: A Challenge to Technology Management*. Int. J. Technology Management. Special issue on strengthening corporate and national competitiveness through technology, 1992, 7: 17-39.

2. M. Callon, and J. Law, *On the Construction of Socio-technical Networks: Content and Context Revisited.* Knowledge and Society: Studies in the Sociology of Science Past and Present, 1989, 8, pp.57-83.
3. M.Cooley, *Architect or Bee*? The Hogarth Press, 1987.
4. P. Davies, *The Cosmic Blueprint*, Unwin paperbacks, 1989.
5. K.S. Gill, *Summary of Human-centred Systems Research in Europe.* NTT Data Communications Systems Crop. Research Institute of System Science, Tokyo, 1990.
6. P.A. Hansson, Chaos: Implications for Forecasting. *Futures*, Jan./Feb. 1991, pp.50-58
7. J. James, and A. Bhalla, *Microelectronics, Flexible Specialisation and Small-scale Industrialisation in the Third World.* World Employment Programme research, Working Paper, WEP 2-22/WP 220; International Labour Office, Geneva, 1991.
8. I. Sachs, Transition Strategies for the 21st Century. *Nature and Resources*, 1992, 28, pp.4-17.
9. Y. Sato, *Towards Designing Human-centred Systems in Japan.* EC-Japan Conference: The Future of Industry in Global Economy, Nov. 11-15, 1991.
10. J.J. Solomon, *Science Policy Trends in Industrially Advanced Countries.* Science and Social Priorities - Perspectives of Science Policy for the 1990s, June 5-7, Czechoslovak Academy of Sciences, Prague, 1990, 28-42, 1990.
11. M.S. Swaminathan, *Biotechnology; Reaching the Unreached.* M.S. Swaminathan Research Foundation, Proceedings No.3, Madras, 1991.
12. UNDP, *The Human Development Report 1991*, Oxford University Press, Oxford, 1991, pp. 28-29.
13. W. Xinhua, Trends Towards Globalisation and Global Think Tank, *Futures,* 1992, 24, 261-267.

[illegible] Cartes Track [illegible]

2. M. Callon, ed., Essay on the Construction of [illegible] Technical Networks [illegible] and Qualitative [illegible] Knowledge that [illegible] [illegible] [illegible] Gleams that and [illegible] 1989 [illegible]

3. M. [illegible] [illegible] Press, 1989.

4. A. [illegible] The [illegible] [illegible] [illegible] 1990.

5. [illegible] Gill, [illegible] [illegible] [illegible] Technological [illegible] [illegible] [illegible] Tokyo, 1990

6. A. [illegible] [illegible] 1991 [illegible]

7. [illegible] [illegible] [illegible] [illegible] Space, [illegible] and [illegible] [illegible]

7. M. [illegible] [illegible] [illegible] Working Paper [illegible] [illegible] Labour Office, [illegible] 1991

8. [illegible] [illegible] for the [illegible] [illegible] 1991 [illegible]

9. [illegible] Sato, [illegible] [illegible] [illegible] Global [illegible] 1991

10. [illegible] Solomon, [illegible] [illegible] [illegible]

12. [illegible] [illegible] [illegible] [illegible] 1990, [illegible]

11. [illegible] [illegible] [illegible] [illegible] [illegible] No.[illegible] [illegible]

12. UNDP, [illegible] [illegible] [illegible] Oxford University Press, 1991 [illegible]

13. W. [illegible] [illegible] [illegible] Global [illegible] [illegible]

Index